Fatemeh Farnaz Arefian ·
Seyed Hossein Iradj Moeini
Editors

Urban Heritage Along the Silk Roads

A Contemporary Reading of Urban Transformation of Historic Cities in the Middle East and Beyond

Editors
Fatemeh Farnaz Arefian
University of Newcastle
Singapore, Singapore

Silk Cities & Bartlett Development
Planning Unit
UCL
London, UK

Seyed Hossein Iradj Moeini
Faculty of Architecture and Urban Planning
Shahid Beheshti University
Tehran, Iran

ISSN 2365-757X ISSN 2365-7588 (electronic)
The Urban Book Series
ISBN 978-3-030-22761-6 ISBN 978-3-030-22762-3 (eBook)
https://doi.org/10.1007/978-3-030-22762-3

© Springer Nature Switzerland AG 2020
This work is subject to copyright. All rights are reserved by the Publisher, whether the whole or part of the material is concerned, specifically the rights of translation, reprinting, reuse of illustrations, recitation, broadcasting, reproduction on microfilms or in any other physical way, and transmission or information storage and retrieval, electronic adaptation, computer software, or by similar or dissimilar methodology now known or hereafter developed.
The use of general descriptive names, registered names, trademarks, service marks, etc. in this publication does not imply, even in the absence of a specific statement, that such names are exempt from the relevant protective laws and regulations and therefore free for general use.
The publisher, the authors and the editors are safe to assume that the advice and information in this book are believed to be true and accurate at the date of publication. Neither the publisher nor the authors or the editors give a warranty, expressed or implied, with respect to the material contained herein or for any errors or omissions that may have been made. The publisher remains neutral with regard to jurisdictional claims in published maps and institutional affiliations.

This Springer imprint is published by the registered company Springer Nature Switzerland AG
The registered company address is: Gewerbestrasse 11, 6330 Cham, Switzerland

Foreword

The story of this book is somehow linked with the story of *Silk Cities*: another milestone in its journey. The idea of having an independent platform to maintain and foster contextual dialogues on urban matters emerged during our first conference *Urban Change in Iran*. The idea was later developed to become *Silk Cities* (www.silkcities.org). As the name implies, the notion of Urban Silk Roads highlights connectivity and exchange of knowledge, concerns and idea, not only by urban managers but the professional and academics and ultimately people living in those territories.

As an independent and bottom-up initiative, *Silk Cities* adapted an exploratory approach on ideas and its activities; simultaneously, it maintained to have yearly smaller scale activities, for example, collaborating with Springer, *Urban Series* for the publication of the book *Urban Change in Iran*, based on selected and updated papers of that conference; an international one-day knowledge exchange workshop and participating in exhibitions and running panel discussions in other international conferences. However, during this time as the founder of *Silk Cities*, I was frequently asked the same question: When is the next conference? It appeared that those whom Silk Cities was engaged with wanted another magnetic get-together… *well! When there is a first step, there should be a second one too…* The answer was the second conference, Silk Cities 2017, based on which this book builds on. The second conference was again generously hosted by DPU at UCL. Researchers and practitioners alike explored how to reconnect population to their urban cultural heritage in the Middle East and Central Asia (http://silk-cities.org/conference2017). The combination of thematic sessions comprising papers accepted by the scientific committee, panel discussions and special sessions by guest speakers acted as a cross-generational tool and created a bridge between the younger and more experienced generations, as well as bringing together academia and practice.

The conference examined dynamics of sociocultural factors that affect lives and discussed architecture, urban spaces, people and the reality of challenge on how to work with the multi-scalar levels of identity. Examinations of comparing the change in the forms of urbanisation focused on the interaction between culture and policy, showing that reconnecting population with urban heritage depends on great minds that bring together a variety of moralities with different social and cultural norms. Critical questions were also raised on who does what to whom, and why? Heritage is about selectivity, power and politics as to asserting local, national and international interests; tensions between authenticity and a depiction of an 'accurate' past; layers of history are removed, while others are highlighted. There exist diverse approaches to rethink heritage and restoration through acknowledging and integrating cultural identity, questioning the relationship between this history and the region's physical and social trajectory, the notion of modern and contemporary heritage, the dynamics between urban heritage and cultural identity, the linkages between old and new urban fabrics and monuments, and regional political factors leading to the core questions of what heritage means, who it is defined by, who that definition benefits and how that definition is represented in urban development practice. The geographic coverage included Afghanistan, Bahrain, Egypt, Iran, Iraq, Mongolia, Pakistan, Palestine, Serbia, Syria, Tajikistan and Uzbekistan. The few selected papers here do not fully reflect the geographic diversity and multidisciplinary nature of discussions, provided through traditional presentations and invited guest speakers.

For the structuring reasons it was important to keep the balance in the geographic diversity by limiting the number of papers on each country, and to select recommended papers regardless of the country of their case studies on the other hand. However, there was an imbalance in the number of submitted/accepted papers across the region that influenced the way this book has been structured, with Iranian-related papers outnumbering others. This was predictable as *Silk Cities* is originated from a conference on contemporary Iranian cities, whereas the second conference had a regional focus in order to foster dialogues on neighbouring countries which share many common issues. We balanced the geographic coverage of the conference, though, inviting guest speakers and panel discussions, found in the post-conference report at http://silk-cities.org/post-conference-report. For the book, though, the answer was to have a 'Zoom in' part, which could meaningfully narrow down all discussions in other parts to a single country and provide an in-depth case study.

This book provides evidence-based multi-perspective examinations of cities in historic landscapes. Part I emphasises the crucial role of open public spaces on linking urban heritage and cultural identity, existing potentials and challenges, as well as emerging ideas. For example, those open spaces around heritage locations in addressing a social challenge for connecting migrant populations to their city and the emerging holistic approaches to fieldwork, with a focus on connecting residents to their own heritage. Part II focuses on governing urban heritage that traditionally relies on local governance, often with established quantitative measures and processes. New methodologies are nevertheless finding their ways to enter the rigid planning systems, such as engaging with locals for dealing with declining neighbourhoods. Similarly, potentials are emerging for addressing ongoing tensions between competing forces for new developments in the buffer zones of historical monuments with central locations in cities. Yet there are challenges in the reality of practice that ideally can be examined and dealt with preferably before the implementation. There is a need for institutional arrangements and change if required, as well as leadership and impartial understanding of the consequences of policy formulations in practice. Part III discusses post-war urban reconstruction. The timely discussions here build a solid ground for multidimensional examination of addressing urban heritage and identity with reconstruction of destructed historic cities, with a specific focus on housing reconstruction and the reconstruction of the city beyond the restoration of its renowned monuments, such as citadels. Part IV is the aforementioned 'Zoom In' on urban heritage in Iran and can be read as a standalone manuscript. It offers new ideas on contemporarisation of traditional housing stock, showing that at least some typologies of traditional houses can be updated and meets our current lifestyle. It also deals with the notion of modern heritage for new developments from one side and the adaptation of new facilitative strategies and digital technologies for smart cities for everyday urban governance in historic cities from the other side. Examinations go beyond buildings and neighbourhoods and include urban qanats which might still have running water. The book will be a resourceful reading for academics and practitioners alike working on the subject matter.

... *And what's next?* On the final day of the *Silk Cities* 2017, a highly engaging and buzzing workshop 'Urban Café' facilitated individual reflections on the future directions for *Silk Cities*. It appeared that coming together more regularly is a magnetic hub for all. Suggestions were summarised running regular conferences, workshops and publications... *so the journey continues...* At the time of this publication, *Silk Cities* 2019 was underway: *Silk Cities International Conference 2019* focusing on *Reconstruction, Recovery and Resilience of Historic Cities and Societies*, being co-organised by *Silk Cities*, University of L'Aquila and UCL, and held in L'Aquila, Italy (http://silk-cities.org/2019-conference-overview). During and after the related thematic session at the second conference, the need for further discussions and more in-depth attention to this urgent matter was highlighted: something the third conference intended to address.

As you would expect, *Silk Cities* is still in the making yet it taps on a collective will of both younger and experienced generations of academics and practitioners who share concerns and experiences in dealing with real-life urban matters of cities in question those who see connectivity and knowledge sharing as a strategic way forward, to provide a means for improving the quality of the *Urban Silk Roads*.

Hope you enjoy the book.

Singapore   Dr. Fatemeh Farnaz Arefian

Acknowledgements

The preparation of this book greatly owes to all individuals and organisations supported and contributed to the second Silk Cities international conference. Entitled *Reconnect Population to Urban Heritage in the Middle East & Central Asia*, it was held at UCL 11–13 July 2017. Organised by *Silk Cities* (www.silkcities.org) and the Bartlett Development Planning Unit (DPU, UCL, www.ucl.ac.uk/bartlett/development). The conference also received academic support from the UCL Institute of Advance Studies, Oxford Brookes University and the Urban Design Group UK.

The conference enjoyed generous support and contribution of distinguished strategic advisors who supported it to set the right direction for the event, scientific committee who reviewed papers and provided feedback, also, guest speakers. They are acknowledged on alphabetical order as follows: Dr. Mansoor Ali, Loughborough University, UK; Dr. Farnaz Arefian, Silk Cities & UCL, UK; Dr. Camillo Boano, UCL, UK; Prof. Yves Cabannes, UCL, UK; Prof. Mohammad Chaichian, Mount Mercy University, USA; Prof. Kate Darian-Smith, University of Melbourne, Australia; Prof. Julio D. Davila, UCL, UK; Prof. Iraj Etessam, Emeritus Professor, University of Tehran, Iran; Dr. Kalliopi Foseki, Institute of Sustainable Heritage, UCL, UK; Dr. Cassidy Johnson, The Bartlett Development Planning Unit, UCL, UK; Prof. Muhammed Kadhem, German Jordanian University, Jordan; Dr. Kayvan Karimi, UCL, UK; Dr. Hassan Karimian, University of Tehran, Iran; Prof. Ramin Keivani, Oxford Brookes University, UK; Mr. Hassan Ali Khan, Habib University, Pakistan; Mr. Masood Khan, Aga Khan Development Network, US; Dr. Luna Khirfan, School of Planning, University of Waterloo, Canada; Dr. Otambek Mastibekov, The Institute of Ismaili Studies, UK; Prof. Ali Modarres, University of Washington Tacoma, USA; Dr. Iradj Moeini, Shahid Beheshti University, Iran; Siamak Moghaddam, UN-Habitat, Iran; Dr. Farhad Mukhtarov, Utrecht University, Netherlands; Mr. Benjamin Henri Mutin, Harvard University, USA; Mr. Babar Mumtaz, DPU Associates, Turkey & Pakistan; Dr. Moe Naraghi, Independent, France; Dr. Elena Paskaleva, Leiden University, Netherlands; Dr. Lucia Patrizia Gunning, UCL, UK; Mr. Geoffrey Payne, GPA, UK; Dr. Shahid

Ahmad Rajput, Comsats Institute of Information Technology, Pakistan; Dr. Rania Raslan, Alexandria University, Egypt; Ms. Judith Ryser, ISOCARP, UK; Prof. Ashraf Salama, University of Strathclyde, UK; and Ms. Anna Soave, UN-Habitat, Iraq.

Undoubtedly, generous support from *Silk Cities* volunteers, namely, Ms. Mahya Fatemi, Ms. Maryam Eftekhar Dadkhah, Ms. Eva Coleman, Ms. Fatemeh Khatami, Mr. Ehsan Fatehifar, Ms. Shirin Hakim and Ms. Joy Zhuoya Chen before and during the conference made it happen, and the assistance of DPU master students during the conference was essential. Last but not least, live traditional music performance from the region (se-Taar and Tonbak) by Mr. Matin Alimadadian and Ms. Shadi Shad that created an unforgettable cultural ambience for the event.

Special thanks to all!

Contents

1 Introduction 1
  Fatemeh Farnaz Arefian and Seyed Hossein Iradj Moeini

Part I Urban Heritage and Cultural Identity

2 Public Open Spaces in Bahrain: Connecting Migrants
  and Urban Heritage in a Transcultural City................... 9
  Wafa Al-Madani and Clare Rishbeth

3 Paradise Extended; Re-examining the Cultural Anchors
  of Historic Pleasure Avenues 21
  Niloofar Razavi

4 Landscape Architecture's Significance in the Restoration
  of Historical Areas: The Case of Old 'Muharraq', Bahrain 31
  Islam El Ghonaimy and Mohamed El Ghonaimy

Part II Governing Urban Heritage

5 The Rise of the Facilitation Approach in Tackling
  Neighbourhood Decline in Tehran 55
  Kaveh Hajialiakbari

6 Rebuilding Tajeel: Strategies to Reverse the Deterioration
  of Cultural Heritage and Loss of Identity of the Historic Quarters
  of Erbil, Kurdistan, Iraq..................................... 75
  Anna Soave and Bozhan Hawizy

7 Silk Production as a Silk Roads Imported Industrial Heritage
  to Europe: The Serbian Example 93
  Milica Kocovic De Santo, Vesna Aleksic and Ljiljana Markovic

Part III Post-War Reconstruction and Urban Heritage

8 Craftsmanship for Reconstruction: Artisans Shaping Syrian Cities 107
M. Wesam Al Asali

9 Place-Identity in Historic Cities; The Case of Post-war Urban Reconstruction in Erbil, Iraq 121
Avar Almukhtar

10 Post-war Restoration of Traditional Houses in Gaza 137
Suheir M. S. Ammar and Nashwa Y. Alramlawi

Part IV Zooming In: Urban Heritage in Iran

11 Traditional House Types Revived and Transformed: A Case Study in Sabzevar, Iran 157
Karin Raith and Hassan Estaji

12 Can Modern Heritage Construct A Sensible Cultural Identity? Iranian Oil Industries and the Practice of Place Making 175
Iradj Moeini and Mojtaba Badiee

13 Evaluation of the Prospective Role of Affordable Housing in Regeneration of Historical Districts of Iranian Cities to Alleviate Socio-spatial Segregation 193
Alireza Vaziri Zadeh

14 Integrative Conservation of Tehran's Oldest Qanat by Employing Historic Urban Landscape Approach 207
Narjes Zivdar and Ameneh Karimian

15 Assessing the Pedestrian Network Conditions in Two Cities: The Cases of Qazvin and Porto 229
Mona Jabbari, Fernando Pereira da Fonseca and Rui António Rodrigues Ramos

16 Challenges of Participatory Urban Design: Suggestions for Socially Rooted Problems in Sang-e Siah, Shiraz 247
Elham Souri, Jahanshah Pakzad and Hooman Foroughmand Arabi

Index 261

About the Editors

Fatemeh Farnaz Arefian has academic and practical background in architecture and urban design, and an interdisciplinary Ph.D. in development planning from the Development Planning Unit, DPU at UCL. She is the founder of Silk Cities initiative for knowledge exchange and research on contextual contemporary challenges of cities in the Middle East and Central Asia. Farnaz is affiliated with the University of Newcastle (UON), Australia, and an honorary research associate at DPU, UCL, UK. She combines academic research and education with extensive practical experience and has a background in leading her own private sector consultancy for delivering large-scale urban design/planning and architectural projects in the Middle East and Europe, including post-disaster urban reconstruction. She is invited and visiting lecturer at universities and workshops. Farnaz has published academic and professional books and papers. Publication examples include *Organising Post Disaster Reconstruction Processes* (Springer 2018) and *Urban Change in Iran* (Springer 2016).

Seyed Hossein Iradj Moeini is a senior lecturer in architecture in Shahid Beheshti University (SBU), Tehran—where he has obtained his MA—and a practising architect in London. Having obtained his Ph.D. from the Bartlett School of Graduate Studies, UCL, on contemporary architecture theory and criticism, he is the author and co-author of numerous papers and two books, has also taught in Yazd, Tehran Azad, and Shariati universities, and worked in a range of Iranian and British practices in a professional capacity. A member of the Royal Institute of British Architects (RIBA) and the Iranian Comparative Arts Circle, he is also an amateur photographer and musician.

Chapter 1
Introduction

Fatemeh Farnaz Arefian and Seyed Hossein Iradj Moeini

Beyond the nostalgic admiration of physical urban heritage, such as bazaars and souks, maze-like neighbourhoods and sensational architectural styles of historic cities, there are many challenges that urban governance systems, practitioners and residents of those cities and neighbourhoods face on linking urban heritage, design, planning and development within urban transformation processes. This region, which is geopolitically referred to as the Middle East, is the home to many ancient settlements and early human endeavours that formed these cities. Urban historic characteristics such as historical city centres still exist in many cities particularly along the historic trade routes—Silk Roads—across the region, besides the contemporary exercises of city formation. However, the urban continuity that once existed across generations in physical and social paradigms has been interrupted in the midst of rapid urbanisation, globalisation and urban economic pressures, in addition to conflicts and destructive natural phenomena. Dealing with such pressing issues in a historic city is more complex than dealing with those in newly built cities and urban areas.

This book contributes to the contextual examination of linking urban heritage, design, planning and development within urban transformation processes in the Middle East and its broader context. The overall approach is based on interpretive and multi-perspective accounts. Authors are either from the region or have lived and worked there with sufficient practical experience, understanding sociopolitical, administrative and economic contexts. They offer insider insights on their discussions and enable a deeper understanding.

This book deals with contemporary cities of the region and pays attention to the '*what about now*' question. It concerns with the correlation between cultural iden-

---

F. F. Arefian (✉)
University of Newcastle, UON & Silk Cities, Singapore, Singapore
e-mail: farnaz.arefian@newcastle.edu.au; f.arefian@silkcities.org

S. H. I. Moeini
Shahid Beheshti University, Tehran, Iran

© Springer Nature Switzerland AG 2020
F. F. Arefian and S. H. I. Moeini (eds.), *Urban Heritage Along the Silk Roads*,
The Urban Book Series, https://doi.org/10.1007/978-3-030-22762-3_1

tities, their respective histories and their relevance to our ever-changing globalising region, which also faces specific contextual challenges, such as destructive conflicts. Cities are complex entities interconnecting economic, social, cultural, physical and administrative systems simultaneously in one place. Cultural heritage, in essence, relates to both physical and social sciences. It connects to identity and is still seen in its concrete manifestation in the physical space, i.e. built environment, as well as in its observance of social values, experience and intangible dimension. Thus, urban cultural heritage relates to urban elements (urban morphology and built form, open and green spaces, urban infrastructure), architectural elements (monuments, buildings) and community rituals and ceremonies, rooted in their histories and values. Historic urban landscape and people's intangible cultural identities are connected and interpreted in the built environment in various ways. Ideally, responsive policies should be operationalised and the urban development should integrate the process of reconciliation between urban heritage and cultural identity in the middle of forces of globalisation and standardisation. In practice, local and municipal governments, which are key parts of any activity on historic urban fabrics and neighbourhoods, morphology, buildings and open spaces, are often under-resourced and ill-structured within quantitative criteria, regulations and processes. While the rhetoric advocates urban heritage and cultural identity to be a driver for city's development that engages with broader population and its residents as opposed to tourist-driven heritage preservation, the question of implementation and practice remains for this region. How can cultural identity and the sense of belonging become a driver for advocating and preserving urban heritage beyond the museum approach and exhibitory objects? How can people's ways of life, customs and ceremonies influence their relation with urban heritage? Furthermore, in addition to the global challenges of developing countries, the region also suffers from conflicts, wars and destructions by earthquakes, landslides, floods and so on. While dealing with post-war reconstruction in their uncertain traumatic aftermath is complex, the existing historic context of the city adds further layers of complexities to the issue. As a result, the region has faced a growing need of reconstructing and retrofitting cities.

The content of this book is based on updated versions of selected contributions to the second *Silk Cities* international conference entitled *Reconnect Population to Urban Heritage in the Middle East & Central Asia*, held at UCL 11–13 July 2017. Organised by Silk Cities (www.silk-cities.org) and the Bartlett Development Planning Unit (DPU, UCL), the event was also academically supported by the UCL Institute of Advanced Studies, Oxford Brookes University and the Urban Design Group. The conference enjoyed generous support and contribution of the distinguished strategic advisors who supported to set the right direction for the event, scientific committee who reviewed papers and provided feedback, also, guest speakers as mentioned before. Accepted papers and guest speakers addressed Afghanistan, Bahrain, Egypt, Iran, Iraq, Mongolia, Pakistan, Portugal, Gaza, Serbia, Syria, Tajikistan and Uzbekistan. Surely, this collection of revised versions of peer-reviewed papers cannot reflect on presentations by guest speakers but it includes insightful discussions which nevertheless reflect on common issues on the subject matter in the region.

To correspond with the multifaceted nature of the subject matter, the contributions are divided into four parts. Part I explores the process of reconciliation between urban heritage and cultural identity amidst forces of globalisation, developmental challenges and standardisation. The overarching focus is on public open spaces through which we experience the city. *Wafa Al Madani* and *Clare Rishbeth*'s contribution looks at potentials of public open spaces in Bahrain, not just to strengthen a sense of attachment to the place for native people, but perhaps more importantly, for migrants. By implication, these spaces can play a key role to bring communities together and avoid a ghettoisation of the city. A lot has been said about migrant communities in cities and the risks of them being confined to ghetto-like areas without any meaningful integration happening between them and their host communities. This chapter makes a case for urban spaces to contribute to the reconciliation of different cultural identities. However, full potentials of public open spaces, are often not realised. Bringing a neglected element, this time in the study and analysis of historic public pleasure walks in the region, *Niloofar Razavi* highlights their social aspects as crucial in such spaces' thriving and survival. Going beyond examples of gardens in their native Iran, they focus on some Timurid Indian examples in search for parallels and come up with suggestions as to how to approach the preservation of these spaces, through which residents and outsiders experience the city. In their contribution about the significance of landscape architecture in restoration projects, *Islam and Mohamed El-Ghonaimy* also criticise the exclusion of landscape concerns from urban restoration projects, and using the case of their work on Old Muharak, Bahrain. They show how such consideration can be introduced into restoration processes. They argue that this can result in a more sustainable practice—socially, economically and environmentally.

Part II focuses on governing urban heritage that traditionally relies on local governance and municipalities, often under-resourced with established quantitative measures and processes. The official international recognition of urban heritage as world heritage sites will usually bring attention to local and national scales and the issuance of new guidelines for the neighbourhood and its buffer zone. However, a responsive implementation process and funding strategy are needed for their deliverability. The state of decline many historic cities suffer from may have its origin in a variety of reasons: from making the city more 'efficient' to trying to eliminate memories of a now out of favour past and so on. Local people might be left away in new plans or their participation might not be carefully programmed. *Kaveh Hajialiakbari*'s contribution focuses on the work of Facilitation Offices in dealing with problems of declining neighbourhoods in central Tehran. Facilitation Offices are neighbourhoods' administrative wing of municipalities to implement participatory planning and design. Whilst he acknowledges the importance of authorities' interventions in dealing with problems of such areas, he also adopts a diagnostic approach by assessing their working mechanisms and their shortcomings. He concludes with a list of such shortcomings, rooted more or less in the Iranian bureaucratic machinery on one hand, and the top-down, physical structure-oriented official views about development among the country's authorities, on the other, as the characteristics which to some extent are common in the region. *Anna Soave* and *Bozhan Hawizy* focus on Erbil, Kurdistan, Iraq and provide other examples of collective international and local efforts in order to stop and 'undo' the deterioration and loss of cultural identity brought about by the recent wholesale property expropriation of the city's historic quarters. They explore potential new innovative urban projects that do not have to rely solely on government funding or large-scale developers, while relying on local socio-economic potential and its legal, financial and environmental constraints. The inclusion of case studies not geographically part of the Middle East region might not seem an obvious choice. But the study by *Milica Kocovic*, *Vesna Aleksic* and *Ljiljana Markovic* on industrial heritage in search of possible alternative ways of governance offers a fascinating insight into linking industrial heritage and governance. Their work on silk industries in the Balkans examines how these industries bring with them not just new forms of horticulture—the planting of nonindigenous white mulberry trees—but an entire mode of cultural interaction and exchange facilitated by the Silk Route.

Part III reflects on the urgent need of some cities in the region, dealing with post-war urban reconstruction and urban heritage, beyond monuments. This part seeks to address how the history of a city, its built environment and its people inform urban reconstruction and retrofit processes, and expose past and present negligence. The separation between architects and builders can be traced back to the Renaissance period in Western countries, when architects' career started to be recognised as the superior one of the two. Focusing on Syria, *Mohammad Wesam Al Asali* shows that the craftsmen guilds structure, among other things, kept construction trade integrated until the much later early twentieth century. He examines how the country's construction industry has suffered from an eventual breakup, with builders having no representation in the system, but actually keeping a large share of developments in their hands, in fact more than architects. The question is how any reconstruction plan could be more inclusive and recognise those builders? *Avar Almukhtar* also looks at Erbil, Kurdistan, Iraq, focusing on housing sector in the newer parts of the city—beyond historic neighbourhoods and buffer zones. He examines the emerging place-identity, gated communities and conflicting identities that are consequences of the post-war urban reconstruction process in Erbil. While the monumental citadel in Erbil is the centre point of attention, as he points out, the city's sense of place goes beyond just there. There is a need for an inclusive comprehensive approach in the planning process to contribute to the overall local place-identity of the city. *Suheir Ammar* and *Nashwa Alramlawi* provide details on how conflicts damage traditional houses, which nevertheless might already have suffered from previous negligence. Highlighting the importance of the active engagement of locals in the process, their paper discusses the post-war reconstruction and retrofit in historic neighbourhoods in Gaza, providing a local perspective on this collaborative work between a local institution and an international organisation.

Part IV zooms in on urban heritage in Iran and examines the notion of contemporarisation on one hand and the adaptation of new strategies and technologies for urban governance on the other. Those strategies and technologies are often rooted in Western approaches and borrowed by the countries in the region—or any other developing country. This part examines how those strategies and technologies work in practice and influence urban life and identity. It starts with chapters dealing with questions of scale, time and attempts to contemporise physical on linking urban heritage and identity. Among justifications for abandoning traditional building types and embracing modernist approaches in designing new developments, particularly in rapidly developing societies, is that traditional types cannot accommodate new lifestyles. In their study on Sabzevar's traditional house types, however, *Karin Raith* and *Hassan Estaji,* come to a different conclusion: that although some types admittedly can no longer be used for new developments, there are certain other types capable of being updated and used for present-day housing design. New houses can, in this way, benefit from the more measured relations which traditional houses have enjoyed, both with their surrounding built environment and with local climatic conditions. Whilst modern developments are often blamed for lacking a sense of place, *Iradj Moeini* and *Mojtaba Badiee* focus on South Iran's modern oil towns built in the earlier days of the industry as a different experience. Although developed by non-locals, authors observe, the sensible approach towards the area's harsh climate and a sense of respect for residents, together with a well-organised management system, have resulted in a sense of place, and in the built environments which have robustly stood the test of time. This raises the question of whether it is modernist development per se or the way it has been delivered elsewhere, which has weakened a sense of attachment to the place. New developments in historic parts of cities have been addressed elsewhere in this volume. *Alireza Vaziri Zadeh,* however, particularly focuses on housing developments in such areas, and how they can be used to improve poor living conditions in such areas in a variety of ways. He warns, however, that such developments can only have positive effects if delivered considerately. He offers a number of suggestions, indeed prerequisites, for this to happen.

The following chapters examine everyday urban development in cities and the adoption of new measures, strategies and technologies by local governance systems for addressing unique urban characteristics which a city has. Not all that characterises an historic city are its buildings. Urban infrastructures have played their part, notably, the water supply systems in dry climates such as that of central Iran. Widely fallen into oblivion, Iranian qanats such as Mehrgerd are no longer considered as a vital part of modern-day water supply system. This qanat, however, as *Narjes Zivdar* and *Ameneh Karimian* have shown, still possesses running water, and in need of serious preservation work. What they propose is that an assessment of the qanat is done based on UNESCO's Historic Urban Landscape approach, which authors find appropriate as a starting point for conservation work in this qanat. We can similarly point at access networks in historic and indeed all cities as a key defining element of their character. Focusing on this aspect, *Mona Jabbari, Fernando Pereira da Fonseca* and *Rui António Rodrigues Ramos* look beyond the

Middle East region to the other—Western—side of the Silk Roads and compare the Iranian case of Qazvin with Porto in Portugal using Space Syntax, and thereby suggest a pedestrian network assessment system whose application may well go beyond these particular historic cities. Contemporary theories and practices advocating participatory urban planning and design methods have been circulating around since suspicions raised about modernist elitism. In their work on an historic, deprived area of Shiraz, *Jahanshah Pakzad, Elham Souri* and *Hooman Foroughmand Araabi* make a case for such methods by showing how they can be put into practice, and how significantly they can improve conventional urban design methods, particularly when it comes to historic deprived areas.

Through above examinations, this book offers multi-perspective insights towards a deeper understanding, unfolding complexities of historic cities and cultural identities by those who not only studied on those cities but also lived and worked there. Those evidence-based discussions set the pathway for improving urban practices and future researches on the contextual issues of contemporary urban development building on a strong historical background, especially those along the historic routes.

Part I
Urban Heritage and Cultural Identity

Chapter 2
Public Open Spaces in Bahrain: Connecting Migrants and Urban Heritage in a Transcultural City

Wafa Al-Madani and Clare Rishbeth

Abstract Developed around a question of what urban heritage might mean for migrant populations, this paper argues that urban heritage sites are inclusive spaces and that protecting heritage does not conflict with transcultural processes and adaptation of migrants. In common with many Silk Roads cities, Bahrain has highly modern urban development alongside ancient heritage sites, with a population profile of more than 50% migrants. The paper sets out selected findings from a landscape architecture PhD research, which explores the role of public open spaces in transcultural cities. This research used a qualitative methodology, integrating observation, on-site short interviews, in-depth go-along interviews and expert interviews. The participants included both Bahraini and migrant groups from different generations. The intention was to understand diverse personal interpretations and socio-spatial associations, and analyse these alongside different patterns of use in urban public spaces. This paper focuses on three sites in particular: Bab Al-Bahrain area, a historic square and souk, the Pearling Trail, a UNESCO Heritage site, and the Hunainiyah Park, located in a desert valley beside a fort. The findings highlight the fact that these heritage sites also act as significant transcultural sites for migrant leisure practices. These sites were connective locations for people and provided support with a sense of belonging and welcome. The evocative quality of these places also prompted memories that connected migrants to different heritage locations and experiences in home countries. Participants valued these sites as places to gather, echoing patterns of socialising that span centuries. We conclude that these heritage sites can be both highly rooting at the local scale and support transcultural adaptation. Public open spaces and landscape character are integral dimensions of heritage, hence landscape architectural practice can play a vital role in achieving the agenda of both conservation and social inclusivity.

W. Al-Madani (✉)
University of Bahrain, Isa Town, Bahrain
e-mail: walmadani@uob.edu.bh; wafaalmadani@hotmail.com

C. Rishbeth
University of Sheffield, Sheffield, UK
e-mail: c.rishbeth@sheffield.ac.uk

© Springer Nature Switzerland AG 2020
F. F. Arefian and S. H. I. Moeini (eds.), *Urban Heritage Along the Silk Roads*,
The Urban Book Series, https://doi.org/10.1007/978-3-030-22762-3_2

Keywords Public opens spaces · Urban heritage · Migration · Transcultural practices · Bahrain

2.1 Migration in the History of Bahrain

Throughout its ancient history, Bahrain, located in the Gulf region, developed a fluid population of multi-ethnic origin from trading and pearling; and later, the oil boom also led to rapid and intense migrations. Bahrain's ideal positioning on the trading marine route between the ancient Mesopotamia and the Indian subcontinent made it an ideal stopover for traders on their voyages, and Bahrain's pearl trade continued to prosper until the nineteenth century. Other goods such as dates and horses were also exported to the world (Al-Rasheed 2005). Trading vessels entered through the old Manama Port and merchants remained for months in order to strike the best bargains.

Several studies have depicted Manama as a cosmopolitan city over the past 150 years and these influences have shaped its social and physical urban fabric (Fuccaro 2005; Alraouf 2010; Ben-Hamouche 2004). Transnational migration resulted in new forms of hybridity that are reflected in the urban framework (Al-Rasheed 2005) and are currently seen in different locations around Bahrain. In the sense of migration, hybridity means cultural mixtures from different geographical locations and over different time periods; hence it represents the history of migration and cultural practices and identities of both migrant and host populations (Hutnyk 2005). Fuccaro (2005) and Onley (2005 and 2014) found that trading and migrant communities in the Gulf shaped everyday practices. Many of these traders made Bahrain their centre for business and moved here with their families, and have become well established in Bahrain's society. Persian and Indian communities are prime examples of this migration history (Fuccaro 2005; Onley 2005). Trading with Iran and India and importing fine quality commodities from shawls, woollen clothes, velvets, silks, shoes, spices, rice, tea, rosewater, liquors and books influenced transnational associations within both the exporting and importing countries (Fuccaro 2005). Moreover, the building of Indian temples and Persian ma'tams in Bahrain indicates the impact of these connections on the urban environments. In the mid-nineteenth century, coffee houses in Bahrain sported a cosmopolitan veneer and differed from those found in the neighbouring Arabic countries (Palgrave 1883). The sustained heritage neighbourhoods with the juxtaposition of churches, ma'tams, mosques and temples are witness to Bahrain's distinguished hybrid heritage.

The discovery of oil in 1932 brought prosperity to Bahrain; and in this context, Ben-Hamouche (2004) claims that with the onset of the oil boom, Bahrain commenced a modernisation drive and adopted an all-inclusive welfare policy, which became a significant force in shaping the country. In addition, Bahrain's egalitarian stance as a country that protects the rights of all people and its innate Arab-Islamic hospitality make migrants feel welcome and led to a population increase mostly among non-Bahraini migrant residents, who now comprised more than half of the population. The economy experienced mass growth, which necessitated the recruitment of both skilled expatriates and unskilled migrant workers due to the shortage of a local workforce.

As described, transnational migration in Bahrain is not new. Though historically migration numbers were less than today, this longstanding history has shaped the hybridity that is integral to the urban fabric of contemporary Bahrain. The contemporary nature of migration has increased diversity with new patterns and with social and cultural dynamics. Dayaratne (2008) asserts that the patterns of migration increase the cosmopolitanism and hybridity in the social fabric, as well as in forms, use of spaces, and elements of architecture using both old and the new in addition to local and foreign. In a context of modernisation and globalisation, the questions are how to balance with the notion of protecting heritage and what urban heritage might mean for this diverse population.

2.2 Urban Heritage, Transcultural Practices and Open Spaces

This chapter frames urban heritage with regard to transcultural practices in public open space (POS). Urban heritage includes cultural practices and life expressions that flow with migrants across borders. Urban open spaces are heritage sites that both reflect cultural identity and where the cultural leisure practices are visible. Life actions post-migration is a process of hybridity and is an integral part of evolving cultures. Hou (2013) links the term transcultural to the twenty-first-century phenomenon of cultural hybridity and fluid identity in an urban context. He further states that the discourse of transcultural cities has emerged as a result of the need for a deeper understanding of today's urban setting within cultural dynamics.

The studies on transcultural cities concern the processes of physical, social and emotional adaptation and hybridity in new places. Migrants' transnational identities are shaped in a hybrid form to enable them to continue practising and enjoying their urban heritage and cultural traditions in urban spaces. From this perspective, the environment has become significant to support adaptation and to build a sense of belonging amongst migrants and newcomers. Rishbeth and Powell (2013) explore the role of the space in triggering memories and forming spatial associations. They conclude that such attachments developed more from the creative engagement between the local and transnational rather than merely from sentimental recollections or nostalgia.

These new concepts of associations in the public realm have supported the development of studies on transcultural cities and can also support studies on urban heritage alongside the forces of globalisation and migration; however, complexities of transculturalism need to be addressed in the practices and theories of urban studies with a greater focus on socio-spatial transformation. This chapter argues that urban heritage sites are inclusive spaces and that protecting heritage does not conflict with transcultural processes and adaptation of migrants. It explores the role of outdoor spaces in continuing the cultural practices of migrants in a new place, and developing a sense of local belonging. Questions here are:

- Do new migrants feel a connection with heritage sites in Bahrain, and in what ways do the cultural practices they bring change the character and social uses of public open space in this country?
- Do these processes challenge a singular normalised understanding of heritage and cultural identity, and how might this require a response in the professional practice of landscape architecture and heritage conservation strategies?

2.3 Data Collection Methods

The empirical research broadly adopted an ethnographic approach and used multiple qualitative data collection methods that are primarily conducted in the field. Within a research program on eight case study areas and involving observation and a total of 98 on-site interviews, both in-depth and short semi-structured, the research focuses on three locations that are relevant to heritage discussions. The fieldwork was conducted in 6 months between July 2014 and January 2015 with a total of 70 observation sessions covering different times, days and seasons. Observations and short interviews revealed different patterns of use in public open spaces. It was made possible to understand different activities at the site through close observation; however, interviews provided additional information about the people's preferences, values and identities. These supported the analysis to determine how different uses of public open spaces are integrating and changing along with people's values, gender, generational and transnational identities.

Thirteen go-along (walking) participant-led interviews were conducted to gather more in-depth data relevant to the affiliation between people and place. This type of interview is a hybrid between observation and interview methods, which allows the researcher to observe participants while, at the same time, access their experiences, reflection and interpretations (Kusenbach 2003; Anderson 2004; Carpiano 2009). Anderson (2004), Jones et al. (2008) and Holton and Riley (2014) find that go-along interviews can stimulate the memory and tell a story through participants' own expressions about their connection to certain places.

The research followed an ongoing sampling process to select the respondents. This way of sampling broadly reflected users of these public open spaces. The sampling was based mainly on people's willingness to be interviewed and their age and language skills. Interviewees included diverse users of Bahraini and non-Bahraini origins.

Expert interviews were also conducted to inform design approaches and policies for planning and managing open spaces. They provided an overview of the current policies and was also an opportunity to support the analysis and to confirm the key issues that emerged from the collected data.

The research analysis was conducted using qualitative data analysis and took some principles from the Grounded Theory Analysis Approach that developed coding plan directly from data. The analysis is facilitated by Nvivo—a computer-assisted qualitative data analysis software.

2.4 Case Study Areas

The three studied urban heritage sites, Bab Al-Bahrain, the Pearling Trail and the Hunainiyah Park, have all received investment with regard to design and management for both restoration and recreation, with the aim that these sites will appeal to both tourists and locals. Bab Al-Bahrain is located in Manama and has a city centre character. It is a congested area, surrounded by high-rise buildings, which also includes the souk and the historical residential area that still features traditional buildings. Bab Al-Bahrain Square is also considered to be the first formal public open space in the region. In its early days, it overlooked the old Manama port. The square's name is derived from the Bab Al-Bahrain heritage monument, which means Bahrain gateway. The souk that is directly accessed from the gate is an Arabic cultural urban space. In the souk, the ubiquitous presence of veteran Jewish, Indian, Arab and Persian traders in currency, gold and other businesses is a distinctive feature. There are several Shiite ma'tams and Sunni mosques in this vicinity along with a 200-year-old Hindu temple located in the heart of the souk. The area has many historical and traditional restaurants, which offer a range of international food; so migrant workers commonly frequent Bab Al-Bahrain to enjoy a meal from their home country. This hybridity, which is part of the history and heritage of the space, provides gathering points for diverse users, particularly from Asia. The roundabout in Bab Al-Bahrain square has a unique appearance, where crowds of Asian migrant male workers congregate (Fig. 2.1).

Hunainiyah Park is located in a desert valley, an important indigenous topography in the landscape of Bahrain. This area also includes a 200-year-old renovated fort and historical neighbourhoods over the ridge overlooking the park (Fig. 2.2). In ancient history, Hunainiyah Valley had a famous well that attracted many people from all over the island seeking its fresh water. The well dried up; but the Hunainiyah Park, which is a well-maintained and designed park, has been implemented in the location as that spring.

The Pearling Trail is a UNESCO World Heritage Site and documents the history of an era when the economy was dependent on the pearl. The Pearling Trail Project is currently undergoing a restoration process as a tourism initiative and integrates elements of contemporary urban design. The project will be provided with a walking path through the location and will create 19 micro-POSs in the sites of some historical houses (Fig. 2.3). During the fieldwork, two of these POSs were implemented. The surrounding area of the selected site includes a new cultural centre, heritage homes, traditional coffee houses and souk.

Fig. 2.1 Bab Al-Bahrain square. *Source* Photo by the Author Al-Madani

Fig. 2.2 Hunainiyah Park. *Source* Photo by the Author Al-Madani

Fig. 2.3 A micro public open space in the Pearling Trail. *Source* Photo by the Author Al-Madani

2.5 POS as Urban Heritage Connective Locations

2.5.1 *Transcultural Sense of Belonging to Urban Heritage*

In order to better understand transcultural connections, the interviews explored migration histories and memories of places at home and in host countries. In the findings, these connections have become part of migrants' social and cultural practices and foster a sense of familiarity and new belonging across borders. An Indian single mother who had moved to work in Bahrain 16 years ago mentioned that Bab Al-Bahrain reminds her of India. She used phrases such as '*Like my India,*' '*I feel here it is our country*' and '*I feel it is my place*', which unmistakably express her intimate affiliation and belonging to the area. Regarding Bab Al-Bahrain square, she added, '*We had in India exactly the same place: a round area with water [fountain], where people were sitting, drinking tea and eating samosa. So, it is exactly like India*'. In another go-along interview, a migrant worker who has been in Bahrain for 2 years, made a reference to a bonding with his homeland when he visits Bab Al-Bahrain, '*There is one place in India like this. It is also a "bazaar". What happens here is similar to the "bazaar" in India, same activities*'. For these two participants, the square site is a hybrid space where 'here' (the Arabic souk) and 'there' (the Asian bazaar) are intertwined. Though they are not residents in Bab Al-Bahrain area, the attachment to the space is informed by repeated visits to the area in which they have become more familiar with the space as '*insiders*' (cf. Armstrong 2004; Powell and Rishbeth 2012).

While a space like Bab Al-Bahrain has a transcultural connotation for Asian migrants, it is not solely a migrant 'ethnoscape' (Irazábal 2011). It also has authentic social values and interpersonal attachments for Bahrainis, particularly for the older generations. While many Bahraini families have moved out of the area, they continue to visit regularly to reconnect with their past and reaffirm their social relationship with the area. A middle-aged Bahraini employer claimed that he drives to Bab Al-Bahrain area every weekend and commented: '*I'm so attached to this "souk"; however, when I talk about it to you or my sons, I can't explain it because this is something that cannot be expressed. The age I lived here and my memories in the space link me to the space*'. This shows a shared sense of belonging to heritage sites amongst migrants and locals.

2.5.2 Patterns of Sociability and Leisure in POS as Urban Heritage Practices

The role and forms of sociability in these historical sites are also part of the heritage and arguably integral to their historical quality, yet is not often considered with professional conservation strategies for historic neighbourhoods (UNESCO 2010). Bahraini values of socialising and hospitality in public spaces within its diverse population give shape to the notion of local culture rooted transnationally in an Arab-Islamic heritage. In Bahrain, it is fairly common to observe people gathering and exchanging acknowledgment with others in front of houses, mosques and shops in their neighbourhoods. In the Pearling Trail site, a Jordanian resident from Muharraq who frequently traverses these spaces on foot mentioned that he feels a sense of local belonging as others often greet him in his way. The middle-aged Bahraini participant during the go-along interview in Bab Al-Bahrain was cordially greeting Indian and Pakistani shopkeepers a few of whom even remembered him from the days of his youth. Those shopkeepers had long-established roots within the neighbourhood and sustained coexistence with the local population. The findings also showed that the traditional coffee houses that scattered along roads were significant heritage and cosmopolitan spaces. These urban spaces hold a distinctive social appeal particularly amongst senior citizens, as men gathering for leisure in these spaces have always been a prevalent feature in Arab cultures. In these spaces, the seating usually spills onto streets and provide visible points for people to socialise. These mundane spaces were also distinctive for migrant foreign workers. The pattern of socialising with diverse people in these historical spaces intertwines with the physical form and fabric of urban heritage.

In parks and gardens, a day out allowed migrant users to be away from their homes for longer and served as a reminder of the leisure activities popular in their countries. Participants expressed their closeness to their homeland when they go in groups on a picnic to a location at a distance with their cooked food. For many Arab migrants, these patterns reflect traces of their desert heritage. Picnicking and gathering in a park help them to reconnect with the desert ambiance through replicating its patterns of leisure and sociability. Besides, in Hunainiyah, a large section of users are Yemenis, and in the words of an old Yemeni grandmother, the fort on the top of the ridge reminds them of Sana'a. Informally defining a group micro-territory is also a common practice in desert places, and is also found in urban settings. In Hunainiyah, it is a familiar sight to find groups of Arab migrant men gathering in the car park beside their cars to chat, have a meal, drink tea or play cards, synonymous to the desert practices. Hence, the cultural leisure practices in parks and gardens could be considered within the agenda of urban heritage.

2.5.3 Complexity of Transcultural Practices

The above case study sites provide a sense of the value of urban heritage for a wide range of urban residents. A key finding is that the local urban heritage spaces have a genuine capacity to be inclusive places in locations shaped by international migration. All three of the selected sites were connective locations that promoted a sense of shared belonging and also supported different patterns of use and adaptation. Using these spaces evokes past memories and often supports the ability for many to adapt leisure practices from other contexts. Sociability and lingering in outdoors spaces should be conceived as integral to notions of urban heritage, and are important forms of inclusion and connection between past, present and future. Hence, the transcultural practices are not necessarily problematic in terms of heritage, instead, they can support each other.

However, it is important to recognise that there are also complexities and tensions related to cultural differences in transcultural cities. While, lingering and socialising outdoors have been leisure and cultural practices over centuries, the fieldwork revealed that the park users were commonly judged by others because of different cultural expectations regarding how POSs should be used. For example, littering, to many people, degrades the idea of public rights; hence, it can be used here to express how people expect others to behave in public spaces. The findings showed that some Arab migrants litter the park due to their perception of freedom, and that they believe it is the responsibility of the maintenance team to keep a place clean. However, these migrant users' behaviours were viewed disparagingly by some other users due to different concepts of cleanliness in POSs. Different parenting practices of some migrant groups in the park are also perceived as irresponsible and lacking duties of care. Bahraini and middle-class participants found it very careless behaviour when some Syrian and Yemeni parents leave their children unaccompanied in the park, while the parents themselves are picnicking and socialising. Another example, hawking during weekends in Hunainiyah Park appeared specifically relevant to migrants' cultural practices. A Yemeni woman in Hunainiyah Park said that she valued her sociability and interaction with other park users while also vending. However, vending activities in the park were judged by Bahrainis who said that there are large crowds of Yemeni visitors who indulge in offensive behaviours like cooking, selling food and hanging up clothes for sale.

Lack of public space management strategies in the professional practice of planning and design when dealing with cultural differences and conflicts can lead to heavy-handed top-down reactions, which lead to exclusion practices and policies that limit the flourishing of different heritages. Hence, a better understanding of transcultural urbanism within the professional practice can be important in supporting inclusivity of urban heritage.

2.6 Implications for Practice: Introducing Transcultural Urbanism to Reconnecting Urban Heritage

In an era of globalisation and migration, the professional practice in the region needs to consider that the historical locations are important sites; not just records of the past, but also to support the continuity of different cultural practices for migrant and non-migrant populations. Parks and gardens need to be specifically considered within the agenda of urban heritage policies, and the landscape architecture can play an important role in reconnecting users to their urban heritage through implementing approaches of transcultural urbanism.

What does a transcultural heritage approach look like in practice?

1. Heritage locations in the region should be accessible to diverse populations and different patterns of use should be facilitated and legitimised whenever possible in these locations. For example, the context of Hunainiyah has a unique heritage parallel to the desert ecology that is valued by many users. The desert has always been considered an elegant natural environment for Arabs. People visit this space repeatedly because it has been provided with a formal park that has many facilities such as picnic pods, benches, sport areas, walkways, play areas, a parking area, toilets and vending activities. The pedestrianised traffic-free urban spaces in the Pearling Trail and Bab Al-Bahrain sites support complexity—different outdoor leisure and cultural practices and patterns of sociability amongst diverse classes and ethnicities. Hence, these spaces support the users' leisure, social and restorative activities in a hybrid form of past and contemporary, elsewhere and here.

2. To ensure social inclusivity and protect diverse heritage practices in transcultural urban spaces, landscape architecture planning, design and management need to recognise that outdoor leisure practices are culturally defined. The research suggests that formal and informal POS (parks, gardens, streets, souks) are important places for these cultural practices. Consequently, standardisation of POS and that the place always stays the same may not be appropriate with social and cultural dynamics in transcultural cities and patterns of use of POSs.

3. Cultural competency should be considered within open spaces and heritage professional practice to incorporate cultural differences and cultural changes through migration. New generations might not use the spaces in similar ways to older generations. Landscape architecture needs to be well informed about these differences and changes. Dialogue and storytelling are effective cultural literacy tools to improve cultural competency and need to be implemented in landscape architecture to understand the nuances of transcultural practices. Within the concept of transcultural urbanism, cultural literacy is effective to understand and mediate the conflicts in urban spaces, which are related to cultural differences. Workshops need to be provided for landscape architects, planners and decision makers to familiarise them with the topic of cultural literacy in dealing with cultural differences and managing conflicts in POS.

2.7 Conclusion

This chapter explored the idea that protecting heritage at odds with transcultural processes and adaptation. The case study locations are important historical places and also appear as significant transcultural sites that support integration. The evocative quality of these places prompted memories that connected migrants to different heritage locations and leisure experiences in their own countries, in which an emotional attachment and sense of local belonging was constructed (even among short-stay migrants). The research also indicates the importance of urban heritage sites and outdoor spaces in supporting migrant experiences as they reflect hybrid cultural influences. Lingering and being outdoors with patterns of sociability appeared in the findings as leisure and cultural practices that flow with migration. Continuing these home leisure practices as forms of intangible urban heritage, in new places promotes well-being for migrants.

The transcultural urbanism approach outlined in this paper sheds a new light on how cultural practices relate to the transformation of urban heritage in open spaces. The findings illustrate the importance of reconnecting to urban heritage as it intertwines with migration and leisure practices in urban public spaces. They findings suggest that continuing home outdoor leisure practices post-migration develops a sense of belonging and promotes wellbeing and a better quality of life. Accordingly, considering social inclusivity and equity in the conservation of urban heritage is crucial. Similarly, the approach of transcultural urbanism appears as a necessary principle to support the reconnection of urban heritage and to respond positively to cultural differences in the region. It is important that landscape theory and practice recognise contemporary migrants' outdoor leisure practices in the conservation of heritage locations. Landscape architecture should also consider different intangible urban heritage practices in planning, design and management of both formal and informal POSs. Heritage can be explicitly inclusive even in times of rapid population changes and that heritage can provide a sense of belonging for everyone, including newcomers to the cities.

References

Al-Rasheed M (2005) Introduction: localizing the transnational and transnationalizing the local. In: Al-Rasheed M (ed) Transnational connections and the Arabian Gulf. Routledge, London, pp 1–18. http://f3.tiera.ru/1/genesis/580-584/…/04957ecf6a63c33a1ad7635cc4f8943b

Alraouf AA (2010) Regenerating urban traditions in Bahrain. Learning from Bab-Al-Bahrain: the authentic fake. J Tourism Cult Change 8(1–2):50–68. https://doi.org/10.1080/14766825.2010.490587. Accessed 13 Dec 2013

Anderson J (2004) Talking whilst walking: a geographical archaeology of knowledge. Area 36 (3):254–261. https://doi.org/10.1111/area.12070. Accessed 19 June 2014

Armstrong H (2004) Making the unfamiliar familiar: research journeys towards understanding migration and place. Landscape Res 29(3):237–260. https://doi.org/10.1080/0142639042000248906. Accessed 15 Nov 2013

Ben-Hamouche M (2004) The changing morphology of the Gulf Cities in the age of globalisation: the case of Bahrain. Habitat Int 28(4):521–540. https://doi.org/10.1016/j.habitatint.2003.10.006. Accessed 19 March 2014

Carpiano RM (2009) Come take a walk with me: the 'go-along' interview as a novel method for studying the implications of place for health and well-being. Health Place 15:263–272. https://doi.org/10.1016/j.healthplace.2008.05.003. Accessed 24 June 2014

Dayaratne R (2008) Vernacular in transition: the traditional and the hybrid architecture of Bahrain. http://uob.academia.edu. Accessed 13 Dec 2013

Fuccaro N (2005) Mapping the transnational community: Persians and the space of the City in Bahrain in C.1869_1937. In: Al-Rasheed M (ed) Transnational connections and the Arabian Gulf. London, Routledge, pp 39–58. http://f3.tiera.ru/1/genesis/580-584/.../04957ecf6a63c33a1ad7635cc4f8943b

Holton M, Riley M (2014) Talking on the move: place-based interviewing with undergraduate students. Area 46(1):59–65. https://doi.org/10.1111/area.12070. Accessed 21 June 2014

Hou J (2013) Your place and/or my place. In: Hou J (ed) Transcultural cities: border-crossing and placemaking. New York, Routledge, pp 1–16

Hutnyk J (2005) Hybridity. Ethnic Racial Stud 28(1):79–102. https://doi.org/10.1080/0141987042000280021

Irazábal C (2011) Ethnoscapes. In: Banerjee T, Loukaitou-Sideris A (eds) Companion to urban design. Routledge, London, pp 552–573. http://www.petronet.ir/documents/10180/2324291/Companion_to_Urban_Design

Jones P, Bunce G, Evans J, Gibbs H, Ricketts Hein J (2008) Exploring space and place with walking interviews. J Res Pract 4:(np)

Kusenbach M (2003) Street phenomenology: the go-along as ethnographic research tool. Ethnography 4(3):455–485. https://doi.org/10.1177/146613810343007. Accessed 31 Oct 2013

Onley J (2005) Transnational merchants in the nineteenth-century Gulf: the case of the Safar family. In: Al-Rasheed M (ed) Transnational connections and the Arabian Gulf. Routledge, London, pp 59–90. http://f3.tiera.ru/1/genesis/580-584/.../04957ecf6a63c33a1ad7635cc4f8943b

Onley J (2014) Indian communities in the Persian Gulf, c. 1500–1947. In: Potter LG (ed) The Persian Gulf in modern times: people, ports, and history. Palgrave Macmillan, New York, pp 231–266

Palgrave WG (1883) Personal narrative of a year's journey through central and eastern Arabia (1862–63). Macmillan and Company, London

Powell M, Rishbeth C (2012) Flexibility in place and meanings of place by first generation migrants. Tijdschrift voor economische en sociale geografie 103(1):69–84. https://doi.org/10.1111/j.1467-9663.2011.00675.x. Accessed 28 Oct 2013

Rishbeth C, Powell M (2013) Place attachment and memory: landscapes of belonging as experienced post-migration. Landscape Res 38 (2):160–178. https://doi.org/10.1080/01426397.2011.642344. Accessed 28 Oct 2013

UNESCO (2010) World heritage and cultural diversity. Germany, UNESCO. https://www.unesco.de/fileadmin/medien/Dokumente/Bibliothek/world_heritage_and_cultural_diversity.pdf

Chapter 3
Paradise Extended; Re-examining the Cultural Anchors of Historic Pleasure Avenues

Niloofar Razavi

Abstract The public pleasure walks in historic urban landscapes, often in form of ceremonial avenues, have had many examples in historic cities throughout the world. Some of the most famous examples of these pleasure avenues have survived in the cities of the Middle East and Central Asia, and have been studied extensively. However, the studies often focus on the corporeal characteristics of these avenues. Notwithstanding the value of such morphological inquiries, they often neglect the properties related to the social dynamics and their impact on the cultural character and mode of survival in these urban features. Therefore, it seems only logical to both deepen and broaden the enquiries to examine the social aspects of historic examples in this region. In hope of revealing the attributes that maintained the cultural character of these urban elements, this article concentrates on social dynamics and spatiotemporal sensitivity of their context. Accordingly, the study proceeds by reviewing the behaviour episodes, modes of interaction, and patterns of visibility for the elite as well as ordinary citizens. The main data is collected from historic documents. The results of the analysis on the historic records and narratives reveals how the changes in social dynamics may alter the use and effect the survival of pleasure walks. Because of the abundance of documents on Chahar-bagh of Isfahan, the study does not focus on this example. Instead, drawing on the successful attributes of Safavid examples, the inquiry tries to find historic predecessors in a Timurid urban heritage in Delhi, India and proceeds to search for the lasting properties in the Qajar examples in Tehran. The chapter suggests that the persistence of this configuration in historic examples may have been the result of specific attributes including but not limited to the planned visibility of a sovereign presence, the deliberate creation of a socially picturesque setting, the aura of eventfulness, etc. It then proceeds to pose the important question that among the many attributes of these urban elements, which ones may be relevant today; which ones are possible to maintain or revive; and most importantly, which ones are considered the 'cultural anchors' and are indispensable in conservation of what was once considered an elongated Persian garden; an extension of paradise.

N. Razavi (✉)
Shahid Beheshti University, Faculty of Architecture and Urban Planning, Tehran, Iran
e-mail: n.s.razavi@gmail.com

Keywords Historic Urban Landscape (HUL) · Pleasure avenues · Cultural anchors · Mughal Delhi · India · Qajar Tehran

3.1 Introduction

Streets and roads, as containers of human movement along a set path, have been found to have definitive impacts on the creation and perception of space. Some researchers believe that the history of street is older than the history of the city (Mehta 2013), since the 'the manner in which the notion of road or street is embedded in human experience suggests that it has references to ideas and patterns of behaviour more archaic than city building' (Rykwert 1978). However, as a discernible part of an urban structure, the remains of first streets are dated 6th millennium BC as the archaeological remains in Khirokitia (Choirokoitia) in Southern Cyprus suggests (Kostof 1991).

The idea of marking both ends of a linear space or path, and delimiting the boundaries along the sides, is not a modern resolution either, and it has been there ever since the *geometric man* and his instruments arrived (Jellicoe and Jellicoe 1998). Therefore, it may come as no surprise that the path to urban or peri-urban gardens has been marked, delineated and elaborated with natural or human-made elements. In fact, they were an elongation of the tree-lined paths within the garden, which were extended to the outside environment to connect the garden entrance to an urban or rural destination. Since the tree-lined path or an *alleé* is 'often the main structural component of a formal garden' (Jacobs et al. 2002), the geometry of their outside extensions is usually just as decisive. Connected to a garden, they also hold the promise of observing the activities of the landlord (royalty or nobles) and the important visitors.

In the Middle East and Central Asia, the historic gardens were presumed to depict Paradise, even before the Islamic era. The etymology of the world connotes an enclosed piece of land, but the symbolism of the heavenly realm has been so strong that paradise is now commonly used to imply heaven. Hence, the road leading to and from such place has often been designed to justify such standing; an extension of paradise.

According to historic documents, these routes often act as a stage for important events too, most of which were either meant for a public audience (Yu et al. 2006), or were attractive enough to entice such audience. As a result, they have usually been the setting for interaction or limited mixing of different social ranks. Accounts show that the configuration of these spaces not only had an impact on the intended or unintended mode of these interactions, but was also affected with the changing of the mode throughout the history. In this respect, historic pleasure walks seem to be a perfect example of how cultural protocols and social interactions are affected by the configurations of a public space, and how the change in cultural aspects might have an impact on the functioning of these urban elements.

In order to examine the initial potential of these elements, and their subsequent changes, this article relies on the historic accounts as well as images and drawings, many of which are actually concentrated on describing the events rather than the setting, and therefore present a colourful interpretation of socio-cultural values and attributes. In addition, because there are narratives from different periods of history, we are provided with a layered, time-sensitive description of the urban space. The diasporic narratives also indicate which aspects had a long-lasting impact on the social dynamics of the space and could, therefore, be considered a 'cultural anchor'. Overall, the preferred mode of inquiry here strongly relies on the Historic Urban Landscape (HUL) approach, which bears the suitable sensitivity to layers of information buried in the tangible and intangible attributes of an urban heritage and its context.

3.2 The Historic Urban Landscape (HUL) Approach

The intricate layering of cultural values in a historic landscape (urban or otherwise), usually poses a methodological issue for heritage studies. Through a search for an all-encompassing mode of inquiry into the life cycles of heritage elements, the most recent methodological attempts mark a shift from concentrating on 'districts' to examining 'landscapes'.

To comprehensively appreciate and recognise the values of urban heritage, there is a need to address all layers of its existence. In order to do so, a search and selection of historic or contemporary theories and practical experiences that would formulate an all-inclusive approach seems to be the logical next step. Such an approach should strive to embrace the tangible and the intangible aspects in both the heritage and its setting. Studies searching for such approaches have found new grounds for argument within the ongoing discourse of 'urban landscape'. The concept of Historic Urban Landscape (HUL) created a definition for the phenomena which consisted of *'ensembles of any group of buildings, structures and open spaces, in their natural and ecological context, including archaeological and paleontological sites, constituting human settlements in an urban environment over a relevant period of time, the cohesion and value of which are recognised from the archaeological, architectural, prehistoric, historic, scientific, aesthetic, socio-cultural or ecological point of view'* (UNESCO 2005). The context-based methodology driven from this definition is known as the Historic Urban Landscape approach.

3.2.1 The HUL Method

Even before the launching of the 2005 HUL approach in heritage conservation management, many attempted to correct the false notions of landscape as a mere physical reality observed from fixed points in space. Their vision holds that only in encoding and decoding the landscape can one understand the collective consciousness that is responsible for creating the end result, since landscape '*represents not only the physical expression but also the hopes and dreams of that culture; an embodiment of their values*' (Motloch 1991, p. 17).

When adopting the HUL approach, we must be aware of the intricate structure of lived experience in a place both today and in the past. One must also consider the fine and often complex threads of relations and values which are added by the presence of singular functions or marginalised communities whose relations are not immediately recognisable. Unless these fine threads are acknowledged, there is a high chance that the dominant view would eliminate the last evidence of other historic realities (Smith 2015, pp. 228–229).

3.2.2 Palimpsest or Brecciation

Long before the HUL approach was adopted as a mode of inquiry and a management tool, the concept of 'palimpsest' dominated urban studies for decades to provide the method and terminology to '*a broadly defined phenomena: the mingling of the past in the present*'. (Bartolini 2014). After the concept was used as a metaphor to evoke the ways in which memories are preserved in form of layers of images and feelings in the 19th century, it found its way into archaeological studies and henceforth to urban studies on the social memory and the city. '*Specifically, the metaphor of the palimpsest is useful to evoke the traces—both material and immaterial—left on the urban landscape and the impact that these might have on citizens*' (ibid.).

However, since the tangible and intangible elements are not composed in an orderly or chronological manner, the relocated or juxtaposed elements are often consolidated in a manner that not only confuses the research, but also complicates the conservation choices. To describe this condition, Bartolini suggests the use of the Freudian expression used to describe the agglomeration of disperse elements that make a dream; brecciation (Bartolini 2014). A piece of breccia is a rock that consists of coarse deposits of sedimentary fragments from different origins that are consolidated or cemented together as a result of intense heating and pressure (ibid.), and is useful when trying to analyse intense collocation of the tangible and the intangible in urban heritage. It is also a useful reminder that in adopting the HUL method, care must be taken not to value all the layers (or all the elements within a layer) equally. It may also explain the endurance of some parts or some attributes of a historic urban landscape and the fading of others. The acknowledgment of the enduring attributes of a heritage, often compiled in a 'brecciation', and the measures taken to sustain them is a point that should be taken seriously in face of apparent neutrality of layering methods.

3.2.3 The Cultural Anchors in HUL

Based on paragraph 82 of the Operational Guidelines for the Implementation of the UNESCO World Heritage Convention (UNESCO 2013), a variety of 'attributes' can be found to embody important cultural values. Along with items such as form, function, design, etc., the list includes '*spirit and feeling*' and '*other internal and external factors*'. Even with all the categorisations and clarifications, the UNESCO admits that some of these attributes '*do not lend themselves easily to practical applications of the conditions of authenticity, but nevertheless are important indicators of character and sense of place, for example, in communities maintaining tradition and cultural continuity*' (UNESCO 2013, p. Par. 83). The World Heritage Operational Guideline states '*spirit and feeling*' as an example, but one may see that '*other internal and external factors*' is just as vague. To clarify, paragraph 84 contends that all these attributes and other '*physical, written, oral, and figurative sources, which make it possible to know the nature, specificities, meaning, and history of the cultural heritage*' are considered information sources, and make the elaborations on the heritage possible.

Based on the points examined above, it is not surprising to find specific points in time or places that have a significant role beyond their time and location in a historic urban landscape. Since their significance is multidimensional and surpasses the layers that we find during the analysis, for the purpose of this study, we identify them as 'anchors'. These anchors are the reminders of the vertical connections between the layers. Connecting these anchors and registering the lines of visible or invisible relations between them creates dynamic time-sensitive layers suitable for this type of investigation. In other words, the cultural anchors are reminders that the HUL method will not be successful in acknowledging all the layers of existence unless we value the intricate interaction of all the layers after the detailed examination of an urban heritage; a careful synthesis after the analysis.

3.3 Cultural Anchors of the Historic Pleasure Avenues

As mentioned earlier, the famous Safavid urban space for public gatherings, namely the Chahar Bagh, is not the focus of this study since years of fascination with this urban element has produced a plethora of research documents, some of which have given in-depth analysis starting from etymology of the name and ending with morphological origins of the form (Ahari 2006). However, for the purpose of this study, finding the cultural anchors that have guaranteed the lasting popularity of pleasure avenues seem to be a proper starting point in the search for strong attributes of these historic elements.

3.3.1 The Eventfulness

There is an argument for the use of these spaces to entice national unity by the Safavid rulers. (Ahari 2011) Apart from the formal ceremonies for the arrival and departure of the court members (especially the king), or welcoming foreign delegations, numerous feasts and recurrent entertainments happened in these urban pleasure areas.

Halbwachs (1980)contended that *'the group not only transforms the space into which it has been inserted, but also yields and adapts to its physical surroundings'* and that *'the group's image of its external milieu and its stable relationships with this environment becomes paramount in the idea it forms of itself'*. Therefore, the profusion of events in these spaces created a collective memory of 'eventfulness' that persists until today, all the time gathering more people and harbouring more events.

3.3.2 The Picturesque Setting

Even though the Safavid dynasty initiated the first Shiite country in the region, it is noteworthy that many of the occasions used to evoke national unity were non-religious feasts celebrated by Shi'a population as well as the ethnic and religious minorities. The accounts of foreign travellers including Don Garcia de Figueroaand Pietro Della Valle contain the description of the many occasions where people from all social groups would take part either as participants or as spectators. In some cases, as in the water-splashing festival,[1] participants were in disguise so that the social rank would not tarnish the rather practical jokes of the ceremony. (Ahari 2011) There are also detailed accounts of how the corporeal configurations of public spaces, including avenues and squares being adapted to accommodate the needs and location of social groups. Chardin (1686) explains that during the Baker Festival[2] the grand Naqsh-e Jahan square was configured to accommodate different groups of Jews, Zoroastrians, Armenians and Indians. The setting of this specific feast also contained a large spectrum of entertainments from acrobatics to wild animals on display. In another occasion, the Red Flower Festival,[3] spectators could enjoy the music and dances performed by teenagers carrying and distributing flowers. Even the serious rituals related to religious occasions, like the Sacrifice Holliday,[4] were accompanied by music and dramatic displays of ceremonial acts (Ahari 2011).

---

[1] Jashn-e Ab Pashan
[2] Jashn-e Shater
[3] Jashn-e Gol-e Sorkh
[4] Eid-e Ghorban/Eid Al-Adha

Overall, the crowd, the rituals and the spatial interventions seem to have been intentionally put together to create a picturesque collective memory. All through the accounts, you would find the attributes acknowledged for the 'picturesque' (Macarthur 2007); the intended chaotic irregularity, the movement, and even the intended disgust in some occasions.

3.3.3 The Sovereign Presence

The presence of the royal court is manifested in both tangible and intangible attributes of the cultural life in these elements, starting with many occasions organised only to observe the activities of the king including his hunting trips, his reception rituals, the election of the court staff, etc. Even the corporeal configuration of Naqsh-e Jahan square and Chahar Bagh Avenue, with frontispieces of different sizes and scales, stands for the idea and possibility of observation by the king or the nobles. In fact, the possibility of presence is such a strong aspect in these spaces that occasions of absence in formal events have been marked in the narratives of both Iranians and foreigners.

It seems that the dynamics of visibility worked both ways; the sovereign graced the ceremony and the crowd with his/her presence, and in turn, was monitored by his/her subjects for the attentiveness and response to assigned duties. Even the usual collocation of gardens and squares in the urban context was a symbol of presence; the garden is a living entity that can never be left unattended, therefore its prosperity stands for the constant presence and attention of the landlord, in most cases the king. Therefore, the frontispiece of most royal gardens where fronted by a square or the vista of a pleasure avenue, which served as the setting for formal events, whether sombre or festive. Presumably, the presence of the sovereign and the observing crowd in these occasions served as a medium of communication that added to the popularity of these spaces.

3.4 Other Examples of Historic Avenues

3.4.1 The Mughal Delhi

In Shahjahanabad (now the inner city of Delhi), Chandni Chowk is an example of public pleasure avenues that leads to the gates of the Red Fort, containing the palaces and the royal gardens on one end, and the Lahore Gate and the other end. Designed by Princess Jahanara Begum, Shah Jahan's favourite daughter in 1650 CE, it literally means Moonlight Square, a name associated with the place because of the grand pool in front of the fort. Apparently, it acted as a mirror for the moonlit sky at nights, and the shops formed a crescent around the pool in front of the gates to the fort.

Chandni Chowk served as a public promenade at the centre of the new city at the time 'as an axis emanating from the emperor's fort' (Mehta 2013). It has witnessed many scenes—happy as well as tragic—the pomp and glory of Mughal times, the plunder and massacre by Nadir Shah's soldiers, the stately royal procession in 1911 (when the Delhi Durbar was held), a bomb thrown at Lord Hardings, the Viceroy, while he was proceeding in state to the Red Fort on December 23, 1912, and the wild tumultuous crowds surging towards the Red Fort to celebrate their independence on 15 August 1947. Apart from these historic associations, Chandni Chowk is justly famous as the commercial centre of Delhi (Mishra 2017).

This avenue was *'the showpiece urban space of the city, tree-lined with a canal running down the centre… with uniformly designed arcaded buildings with shops selling jewellery, sweets, and other wares, with a scattering of coffeehouse'*. (Mehta 2013)

Apart from the dominant presence of the fort, the pool and the crescent on one side, and the Lahore gate on the other, the morphological configuration of the body of the avenue was the result of identical shops in the front façade, backed by Haveli's (grand courtyard houses) and Kuchas (a zone with houses whose owners shared some common attribute, usually their occupation).

'As a primary unit of the urban fabric, the haveli or mansions of the old city offer a window into the city itself. Like the city, the mansions—fragmented, commercialised and rebuilt—remained vibrant even in their dilapidation' (Hasagrahar 2001). The shifting of power in colonial times, as well as the changing status of the Havelis, has changed the character of Chandni Chowk. That is why today it is referred to as a vibrant and crowded commercial avenue rather than a pleasure avenue.

3.4.2 The Qajar Tehran

Some scholars believe that, unlike the Safavid dynasty, the Qajar rulers never used urban ceremonies as a tool to strengthen national unity. Therefore, their commissioned public spaces in the urban setting were not composed for such results either (Ahari 2011). However, since public gatherings were inevitable, we have documented accounts of their happening in different locales all over the city. For example, the Sacrifice Feast seems to have happened in different squares at different times, with the king being present in some and absent in others. (ibid.) In any case, even the unsuccessful examples of public urban space, which did not have the cultural anchor to survive until today, may serve as a learning practice for the purpose of this study.

As for the sovereign presence, there seems to be controversy in the royal behaviour regarding this issue. They seem to have had a colourful presence in some events to receive their subjects, but seem to have established a routine of shooing away the crowd (even by force and batons) whenever the king and his entourage were arriving or departing on a trip, while the music band would play in tower at the gate to the royal avenue to the inner quarters! (Ahari 2011).

Many of the Qajar public spaces in Tehran seem to carry the proper attributes we have so far considered for pleasure avenues. Most of these places, including Lalehzar Avenue (Ahari 2011), Toopkhaneh Square (Mohammad Zadeh Mehr 2002), or Baharestan Square (Ahari and Habibi 2015) have been analysed as for their potential to accommodate social events. Among these, Lalehzar has always been considered the ultimate cultural alleé in Tehran, since it was built upon the impact Champs Elyseé had on the travelling king; Naser-eddin Shah. However, it lacks some definitive attributes mentioned earlier; it did not lead to any distinctive urban element that could carry a symbolic meaning or political importance. More specifically, it did not lead to a garden. Actually, it was built over the demolished rim of a magnificent garden to make way for the whim a king! Nevertheless, it survived until the middle of the next dynasty, and served as the perfect setting for night-time entertainments, including theatres (Ahari 2008). Baharestan and Toopkhaneh also seem to show some signs of the attributes found behind the sustainability of Pleasure avenues, albeit in the form of squares. They both accommodated important events in presence of the king, they both were picturesque enough in the configuration of the audience as well as the dramatic configuration of architectural elements, and they were both extremely eventful especially at the dawn of the Iranian constitutional revolution in 1901 and ever since.

But perhaps the most important example for the purpose of our discussion is the avenue leading to the inner quarters of the court, which was originally called the Almasieh Street, and later came to be known as the Bab-e Homayoun (the auspicious gateway). It stretched between Toopkhaneh and the elaborate gate to the inner quarters of court and the treasury. Like many other Qajar streets, it was lined with trees and lit at night, and the music band playing there in the afternoon marked the setting of the sun. The political importance of this avenue was such that the gateway was demolished with the fall of the Qajar dynasty, with pieces of it then transferred to another location which never provided the same associations, nor did in front onto the same extended vista. As for the avenue, devoid of all the other attributes of a pleasure avenue, it now serves as the commercial hub for the uneventful trading of papers and suits!

3.5 Conclusion: Re-anchoring the Paradise Extension

The survival of historic pleasure avenues and their continued vivacity seem to be the result of cultural anchoring of these spaces to attributes that include the eventfulness, the picturesque setting and the presence of a sovereign. Examples show that the weakening of these attributes, or the absence of spatial configurations that guaranteed these attributes, have resulted in the fading of vivacity and popularity of public pleasure walks. Moreover, the historic documentation shows that the formal gardens in the urban landscape of historic cities served as a symbol of 'presence'; as a depiction of paradise, they deserved to house the elaborate lifestyle associated with the 'promised land', and as such, was a space that was never unattended. Any change in the status of the garden, or the configuration of its ante-space, resulted in a decline in the tangible and intangible aspects of the pleasure walks leading to the gates of the presumed earthly 'paradise'.

In addition, the multifunctionality and eventfulness of pleasure avenues seem to support the mixing of social ranks; a dynamic virtuous cycle in which each of the elements strengthens the other. Put together, albeit in a deliberately chaotic configuration of social interactions, they safeguard the survival of these heritage elements, be as it may, in form of a picturesque presentation that marks the purgatory at the gates of paradise!

References

Ahari Z (2006) The Chahar Bagh Street of Isfahan; a new concept in urban spaces. Golestan-e Honar, Issue 5:48–59 (in Persian/Farsi)
Ahari Z (2008) Lalehzar; the setting for recreation, from garden into street. Fine Arts, Issue 34:5–16 (in Persian/Farsi)
Ahari Z (2011) City, feast, memory; contemplating the relation between the spaces and urban feasts during the Safavid and Qajar period. Fine Arts, Issue 47:5–16 (in Persian/Farsi)
Ahari Z, Habibi M (2015) Meydan-e Baharestan. The Cultural Research Beaureau, Tehran (in Persian/Farsi)
Bartolini N (2014) Critical urban heritage: from palimpsest to brecciation. Int J Heritage Stud 20(5):519–533
Halbwachs M (1980) The collective memory. Harper & Row Books, New York
Hasagrahar J (2001) Mansions to margins modernity and the domestic landscapes of historic Delhi, 1847–1910. J Soc Architect Historians 60(1):26–45
Jacobs AB, MacDonald E, Rofe Y (2002) The Boulevard Book. MIT Press, Cambridge, MA
Jellicoe J, Jellicoe S (1998) The landscape of man, 3rd edn. Thames and Hudson, London
Kostof S (1991) The city shaped. Thames and Hudson, London
Macarthur J (2007) The Picturesque; architecture, disgust and other irregularities. Routledge, New York
Mehta V (2013) The street; a quintessential social public space. Routledge, New York
Mishra A (2017) Important India. http://www.importantindia.com/11236/historicalimportanceof chandnichowk/. Accessed 19 May 2017
Mohammad Zadeh Mehr F (2002) Meydan-e Toopkhaneh. Ministry of Housing, Tehran
Motloch JL (1991) Introduction to landscape design. Van Nostrand Reinhold, New York
Rykwert J (1978) The street; the use of its history. In: On streets. MIT Press, Cambridge, MA
Smith J (2015) Civic engagement tools for urban conservation. In: Bandarin F, van Oers R (eds) Reconnecting the City; the historic urban landscape approach and the future of urban heritage. Wiley, West Sussex, pp 221–248
UNESCO (2005) Vienna Memorandum. s.l.:s.n
UNESCO (2013) Operational guidelines for the implementation of the world heritage convention. UNESCO World Heritage Center, Paris
Yu K, Li D, Li N (2006) The evolution of greenways in China. Landscape Urban Plann 76:223–239

Chapter 4
Landscape Architecture's Significance in the Restoration of Historical Areas: The Case of Old 'Muharraq', Bahrain

Islam El Ghonaimy and Mohamed El Ghonaimy

Abstract For a long time in most Arab countries, and while dealing with restoration projects, the consideration of landscape architecture (LA) was limited. The term 'landscape architecture' was used in a shallow way just for describing pavements and greenery. Recently, however, there has been an increasing recognition of the importance of LA in improving and enhancing the surroundings of such projects. In many restoration projects that considered LA in its plans, worthwhile results are detected. However, not all achieved such results due to a misunderstanding of LA concepts and meanings. In Bahrain, conservation and restoration projects in Old Muharraq partially incorporated LA aspects, with users (local residences and visitors) expressing their satisfaction about their heritage treasure being recovered. However, the proper development of outdoor spaces was ignored in some respects. The sense of history was unclear in the modern design of pavements, lighting, street furniture and so on. In response, the present study aims to propose a matrix to assess LA in restoration of historical areas and to also help with a qualitative evaluation. This objective will be discussed with a focus on sustainability (social, economic and environmental aspects).

Keywords Landscape architecture · Urban discourse · Public opens · Historical areas

---

I. E. Ghonaimy (✉)
University of Bahrain, Bahrain, Saudi Arabia
e-mail: eslam_elghonaimy@yahoo.com

M. E. Ghonaimy
Newcastle University, Newcastle upon Tyne, UK
e-mail: arch.mohamedislam@gmail.com

© Springer Nature Switzerland AG 2020
F. F. Arefian and S. H. I. Moeini (eds.), *Urban Heritage Along the Silk Roads*,
The Urban Book Series, https://doi.org/10.1007/978-3-030-22762-3_4

4.1 Introduction

Most Arab countries are known for their long negligence of discourse regarding outdoor design. Recently, there has been a rising awareness of the importance of landscape architecture (LA) in general and the restoration of historical built environments in particular. Several queries have, therefore, come to the surface in Bahrain, once the significance of LA was realised in urban restoration projects in historical areas.

As a significant aspect of urban conservation projects, LA's interpretation, objectives, principles and the roles it plays in improving the standards of living—and thereby the built environment in general—must be defined. The neglect of outdoors design surrounding the historical buildings has adversely affected these areas. The historical ambience was not recognisable due to haphazard use of modern materials in pavements, lighting, street furniture and so on in these areas.

Thus, the question was whether we have appropriate indicators to evaluate LA in restoration projects in the historical areas. The answer was no. Such evaluation method is still not clearly identified and the awareness of appropriate guidelines for developing the outdoor spaces is still poor. Therefore, clear guidelines for designing the LA in historical buildings were to be sought through an evaluation matrix. This evaluation matrix is a method supported by interviews with the authorities, representatives and users (residents, tourists and experts) (Fig. 4.1). The matrix is based on sustainability themes (social, economic and environmental), and the study takes place in the old residential area there in the historical areas of Muharraq which is one of the most successful conservation projects in Bahrain (Fig. 4.2). This chapter is based on a research that is comprised of the following stages:

- A description of LA objectives in historical restoration projects.
- An introduction of Muharraq, its main features and the pattern of its open spaces.
- Interviews with users, authorities, representatives and experts.
- Analysis using the 'evaluation matrix'.
- Recommendations to help with improving open spaces in the historical area.

4.2 Interpretation of 'Landscape Architecture' in Historical Areas

Generally speaking, open space design is comprised of nature, culture and self-expression as its main elements. The design of landscape includes both 'hardscape' (pavements, landmarks, fountains, light structures, lighting, signs, etc.), 'softscape' (vegetation, trees, shrubs, etc.) and water. The ideal landscape is a balanced blend of soft and hardscape, with improving the quality of social and cultural life in sustainable manner among its main objectives. LA is a

Fig. 4.1 Interviews with ministry of culture, authorities representative and users (residences, tourists and expertise)

Fig. 4.2 The area of the study, old residential area in the historical areas of Muharraq (Planning 2005)

multidisciplinary field, incorporating aspects of the fine arts, architecture, industrial design, environmental science, geology, geography and ecology, with direct effects on the site's archaeology (Fig. 4.3).

According to Murphy (2016), 'as a discipline landscape architecture is situated at the interface of the arts, the sciences, and the humanities'. Even if in a simplistic manner, in archaeological projects, LA encompasses all of the aforementioned fields, in order to afford comprehensive image and homogenise design for outdoor

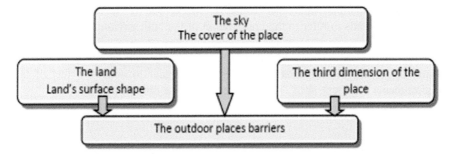

Fig. 4.3 Landscape architecture bounders

spaces. It creates a habitable, sociable, yet equidistantly, aesthetically and visually appealing interactive environment as well.

The statement above is further supported by a claim by Holden and Liversedge (2014) who states that LA 'has developed, to meet the demands of the residents', fortifying the sustainability aspect. It incorporates issues such as climate change and biodiversity. In archaeological projects, it accounts for visual matters between heritage buildings and modernisation process. The built archaeological environment implicates substantial 'open spaces', which are indispensable to heritage buildings, and these spaces are to be designed by a landscape architect giving heed to the quality of the archaeological environment (Gareth Doherty 2016).

4.2.1 The Contribution of Designing Landscape Architecture in Restoring Historical Projects

The *major* emphasis on the contribution of designing landscape architecture in restoring historical projects concerns three main axes as follows:

1. Human behaviour and resources,
2. The built historical environment and natural resources, and
3. Government strategy and urban management.

These are considered as the startings of evaluating the success of L.A. in the pilot study area.

4.2.2 Conservation Project

Conservation and preservation projects seek to preserve, conserve and protect buildings, objects, LA or other artefacts of historical significance. This discipline and its sub-disciplines may be entitled as urban conservation, landscape preservation, built environment conservation, built heritage conservation, object conservation and immovable object conservation. As used by practitioners, 'historic preservation' tends to refer to the preservation of the built environment (Maryland 1997).

Some clarifications are needed before discussing the role of LA in enhancing historical areas. This information enables the designer to acquire a feel of the environments in urgent need for preservation and the urban context.

- Recognising the political, social and economic aspects of the construction period.
- The history of the founder and the events that led to the realisation of that work.
- The chronological study of the events, which and their impacts on the building.

Consequently, information-related matters could be classified into:

- The collection of information, which are inherently placed on the building components with their surrounding context and analysis of architectural drawings.
- The information that could be obtained extrinsically, which are numerous, such as determination of the functions of the building, methodology of utilisations, architectural and technical supplements through manuscripts, documents and pictures.

Therefore, it is essential to consider taking the above seriously in order to create the outstanding LA design for the urban archaeological context in historical buildings which makes them suitable and compatible with the area (Geddes and others 1965).

4.2.3 Fundamentals in Designing Landscape Architecture Within Restoration Projects

The art and science of arranging LA elements under a theme of sustainability are at the core of such restoration projects within an urban context in historical areas (Eldardiry 2013). The spatial allocation for LA elements needs to satisfy the human requirements and demands within the restoration projects in analogous terms with protecting or enhancing the urban environment. Factors such as time, place and activities were hailed as major elements influencing the qualitative performance of designing LA within voids spaces, as follows:

- *Time*: Time is reflected in the places and areas of change, affecting understanding the psychological attitude of the attendees of these areas. Time change indicates the success or failure of the performance and functionality of the site design fields.
- *Movements and travels*: The way the residents and visitors (tourists) move reflects their purposes, whether for shopping, viewing, hiking, gathering,

relaxing or using transport, and in all the previous cases, visual perception and different impression change according to the method of movement and its goals.
- *Places of activities*: Different activities within spaces and places require their associated time and movements, which can have an impact on the environment of historical areas that may affect the visual characteristics and quality (Abdel-Rahman 1991) (Fig. 4.4).

Fig. 4.4 Time, movements and place of activities as major elements to influence the qualitative performance

4.2.4 Influential Factors on the Nature of the Location

LA designs in restoration projects have to be comprehensive. This indisputably involves a systematic investigation of social, ecological and geological conditions. The required scope includes urban design, site planning, town or urban planning, environmental restoration, parks, recreation planning, street furniture, visual resource management, green infrastructure planning and provision, private estate and residence landscape master planning and design. In other words, designing LA projects are not just about softscape and hardscape but it considers all the above items, at both design and evaluation stages.

Within the study, the impetus for cultural preservation items is considered; as the springboard of the need for better understanding across cultures, which has never been greater nor more pressing. The issue is all about the cognitive faculty of the value of heritage subjects and safeguarding them. Restoring the legacy of the former civilisations calls for cultural initiatives, generating creativity, imagination, tolerance, understanding and wisdom well beyond the ordinary (Khan 2005). Thereof, the synergistic effects, a reciprocal relation among LA and the success of urban restoration projects is the prime focus of this research (Table 4.1 and Fig. 4.5).

Table 4.1 Influences on landscape architecture for restoration of historical projects

Factor	Description
Physical (location)	Represents the maximum number of beneficiaries for a known period in a place causing no damage to the environment
Environmental (context)	Use a tolerable level causing no destruction of the environment
Societal and cognitive (awareness for users)	Represented by the costumes, tradition and social conventions that characterise the residents of the place (sociological psychological approaches)
Economic (returns)	Account for the level of use and consumption of place for a material return

Fig. 4.5 Various applications of landscape architecture in old part of cities (historical part)

4.2.5 Methodology for Analysing the Pilot Study

Figure 4.6 shows the methodology of analysing the pilot study and following by the 'evaluation matrix' to find out the missing point and determine the impact of miss apply the LA and the following negative impacts.

4.3 Case Study: Muharraq Settlement, Kingdom of Bahrain

4.3.1 Location of the Historical Part of Muharraq

The Kingdom of Bahrain is distinguished for its numerous historical areas restoration projects. This research project tackles the consolidation of efforts in the restoration of the old residential areas in Muharraq Island as a case study. Muharraq is selected along with seven other cities in the Arab World that has dramatically developed in the past decade, which followed and successfully met the UN Millennium Development Goals (Fig. 4.6).

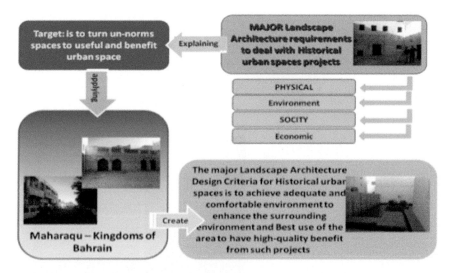

Fig. 4.6 Methodology for analysing the pilot study

4.3.2 Urban Development of the Historical Part of Muharraq

Throughout history Muharraq has significantly influenced the development of the country. It is home to numerous modern economic plants, handicrafts and traditional industries in the Kingdom. As the Kingdom's historical capital city consists of historical fortresses and lavish traditional buildings, it also endows a cultural impact (Fig. 4.7).

In the past decade, the city has witnessed a progressive process, to revive the heritage, regenerate its economic areas, execute a comprehensive infrastructure and shield its communities. The birth of Muharraq was initiated in 1810, with Abdullah Al Fateh establishing himself there as the ruler; organised around a tribal feudal system (Fig. 4.8).

The 2010 population of Muharraq was 189,114, representing 14.3% of the Kingdoms total population with a 7.3% annual population growth. The Kingdom has gone through a series of land reclamations. The area was 757.5 km^2 in 2009 and was in form of an archipelago. The area is expected to reach 973 km^2 in 2030. The area of Muharraq province was 57.46 km^2 in 2009, almost 7.59% of the kingdom's whole area. This area, which used to be less than 20 km^2 in 1976, is annually increasing due to land reclamation. National Strategic Structural Plan for Urban Development NDPS in the determination of large areas of reserved lands intended for Heritage and Ruins purposes (Planning 2005) (Fig. 4.9).

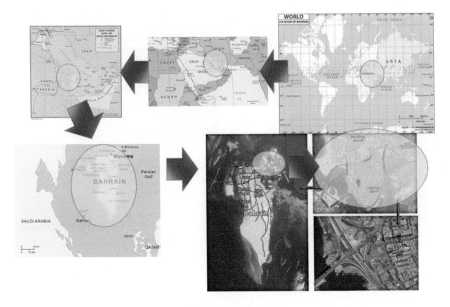

Fig. 4.7 Muharraq Island, Kingdom of Bahrain

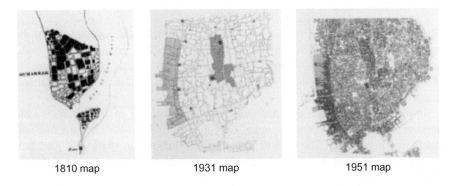

Fig. 4.8 Muharraq Island development

4.3.3 The Residential Pilot Study

The pilot study area is located between the streets Sheikh Abdulla to south, Sheikh Isa to east, Wali Al Ahd and Road No 931 west. The area includes many historic buildings such as Sheikh Hamad ben Khalefa. There has been many restoration projects for many buildings in the area. It has many open spaces 'Baraha' and it is located within Bahrain's touristic map (Figs. 4.10 and 4.11).

Fig. 4.9 Pilot study location within the old residential area in historical part of Muharraq (Planning 2005)

Fig. 4.10 Physical limitation of the study area

4.3.4 Factors Affecting Landscape Architecture in the Historical District of 'Muharraq'

The study of the major factors affecting LA in the historical district of 'Muharraq' was branched into two overriding categories: constant and the variable. The constants are derived from designing outdoor landscape for the historical urban spaces

Fig. 4.11 Photo for the main road and the bonded of the pilot study area

in the city, varied according to proximities, customs/climate economics, and technicalities (El-Ghonaimy 2011a, b) (Fig. 4.12).

4.3.5 Constants Factors

Major constants forces are:

- *Location*: the position of the Muharraq city site coping with the surrounding effects of hot humid climate and desert. Therefore, warm climate and desert characteristics collate, to form the way people live during the evolution of the historical urban spaces process.
- *Ideologies*: Arab cities are ruled by ideologies (beliefs), which are grounded on backgrounds, costumes and traditions, faith and individual values. These

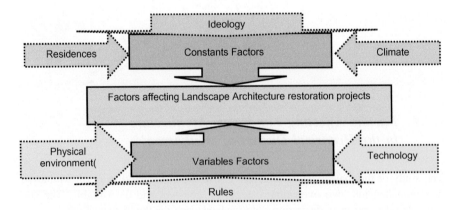

Fig. 4.12 Factors affecting the landscape architecture in the historical district of 'Muharraq'

premises do not undergo rapid changes. It can only be dealt with observing environmental, cultural, social and personal considerations.
- *Rules and legislations*: these constants take long time to be modified (White 1991).

4.3.6 Variable Factors

Variable elements that affect human settlements are:

- *Technical*: it may undergo some changes during a course of time.
- *Nature of the environment*: the environmental equilibrium means that the elements of environment interact according to an ecosystem.
- *Resident (Human) factor*: it is to respect residents' perspective regarding the way they adapt to deal with historical urban spaces. Residents should be trained to understand the value of historical buildings and the ambience of the historical spirit in the area (Al-Abdullah Mohammed 1998).
- *Man-made (built environment)*: the contribution of designers that based on their concepts of the area. It may vary from place to place but generally follows legislations, urban roles and law in designing (El-Ghonaimy 2011a, b).

4.4 Applying L.A. in the Pilot Study Area

4.4.1 'Evaluation Matrix' for Assessing the Satisfying of Landscape Architecture

Based on above mentioned factors of using LA in historical Muharraq a framework/ Main Frame to assist with the case study analysis. It was followed by identifying major problems in outer spaces. This framework respected the vision of sustainability. It is as follows: (Fig. 4.13).

LA design elements can be summed up as follows:

- Hardscape (in terms of pavements, landmark, fountains, light structure, lighting, signs, etc.) and
- Softscape (in terms of vegetation; trees, shrubs, groundcover, etc.) and soft water body.

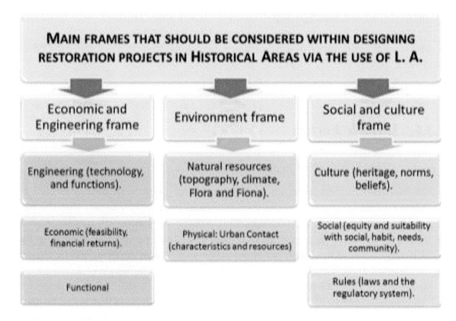

Fig. 4.13 Landscape architecture has to respect sustainability frames

The 'evaluation matrix' (Tables 4.2, 4.3 and 4.4) can be used to identify and evaluate the design of open spaces and their elements. It comprises of two steps:

1) identifying the use or otherwise of a certain LA element:

   ✓: (applied)
   ✗: (not applied)

2) user satisfaction assessment:

| ⬇ Positive | ▲ Highly positive impacts | ■ Average positive impacts |
| ⬇ Negative | ▼ Highly negative impacts | ☐ Average negative impacts |

4.4.2 The Appropriate Use of LA in Outdoor Spaces in the Study Area

4.4.2.1 Engineering and Economic Malfunctions

- Engineering: infrastructure deterioration have negative impacts on traffic roads and their efficiency, exacerbated by capacity restrictions, building maintenance and management issues and inappropriate street furniture such as suitable materials for noise control (Fig. 4.14).

Table 4.2 Measuring the impact of using the LA elements upon the economic and engineering aspects in the study area

Main consideration	Sub	Item	Points	Evaluation	
				Achieving	Satisfying
Economic and engineering	Engineering				
		Technology	In lighting, water body, infrastructure, lighting	✓	▲
		Information technology		✗	
		Building database (use geographic information system (G.I.S.)		✗	
		Suitability of Hard escape materials	Pavements	✓	▼
			Street furniture	✓	☐
			Landmarks	✗	
		Functions		✓	☐
		Safety in shape and function		✓	▲
	Economic			✓	
		Feasibility		✓	▲
		Financial		✓	▲
		Maintenance costs and facilities		✓	▲
		Benefit returns		✓	▲

Table 4.3 Measuring the impact of using the LA elements upon the environmental aspects in the study area

Main consideration	Sub	Item	Points	Evaluation	
				Achieving	Satisfying
Environmental	Environment				
		Nature resources	Dealing with hydro aspects	✓	☐
		Topography		✗	
		Climate		✓	■
		Flora (Soft escape)	Choosing plants suiting site conditions and faculties	✓	▼
		Fauna	Consider feeding and drinking suitable places	✗	

(continued)

Table 4.3 (continued)

Main consideration	Sub	Item	Points	Evaluation	
				Achieving	Satisfying
		Sustainable materials for each design	Colour, feature, texture, maintenance	✓	▲
			Lighting, waste disposals contemnors, light structure, flower bed, trees boxes	✓	▼
	Physical			✓	☐
		Urban contact characteristics		✗	
		Urban contact resources	Time of movements and travels	✓	■
			Places for tourists	✓	▼
			Places for visitors	✗	
			Places for residencies	✓	▲
			Visual impacts	✓	▼

- Economic (feasibility, financial returns): city downtown economic activities are affected by the deteriorated infrastructure conditions, decreased economic and business activities (El-Ghonaimy 2009) (Fig. 4.15).
- Function: most functional landscape elements are missing, limited or dysfunctional (Fig. 4.16).
- There is a lack of connection between the urban landscape and current international economic changes. This project should seek long term benefits not short term temporary ones.
- A lack of private sector engagement and investors to take part in these venture projects along with the government in developing study area that would overcome the large deficiency in the communication grid on the different levels (Eldardiry 2002).
- Deficiencies of the law making system in supporting the development and appropriately controlling it (El-Ghonaimy 2005).
- An effective zoning policy needs to be formulated and adopted to minimise problems caused by mixed land-uses.
- There is a need for applying economic theories in constructing a new approach for urban conservation projects, for example, in terms of 'Build, Operate and Renew (B.O.R.)' to increase its efficiency.

Table 4.4 Measuring the impact of using the LA elements upon the economic and engineering aspects in the study area

Main consideration	Sub	Item	Points	Evaluation	
				Achieving	Satisfying
Social and culture	Social				
		Public participation		✓	☐
		Developing human well being		✗	☐
		Operating management		✓	☐
		Equity		✓	☐
		Suitability with social		✗	☐
		Habit		✗	☐
		Community needs		✓	☐
	Culture				
		Heritage	How can u use these elements to give tooling of proud of your place	✓	☐
		Identity		✗	
		Belongs		✗	
		Norms		✓	☐
		Beliefs		✓	☐
	Rules				
		Thought		✓	☐
		Laws		✓	☐
		Regulatory system		✓	☐
		Sharing private sector		✓	☐

Fig. 4.14 Missing of using LA to cover the engineering facilities

a. The only cafeteria within the area

b. Miss use of facades

Fig. 4.15 a, b Economic activities are affected by the deteriorated infrastructure, and decreased economic activity is the result of neglecting historical areas in Bahrain

Fig. 4.16 Haphazard factions of open spaces

Fig. 4.17 Missing the greenery as L.A. elements in improving pilot study

Fig. 4.18 Courtyards within bigger houses with limited greenery

4.4.3 Environmental Deformities

Environmental deformities are as follows:

- Natural features: in general there is a shortage of vegetation in different levels. Hardscape elements are the major elements used the area. The green vision and eco-friendly materials while applying the L.A. elements are missing (Figs. 4.17 and 4.18).
- Physical urban elements (characteristics and resources); they are represented in the inappropriate urban pattern and spaces categories.
- In general, a mass plan for LA is missing which is a significant part for urban restoration projects, and counted as a considerable factor in dealing with scattered vacant areas and land-use policy, to select suitable sites for services provision.
- Problems resulted from the miss use of lands and narrow roads.
- Hazards resulted from the passing and parking cars which are inside the area.

4.4.4 Social and Cultural Imperfections

- Most residents within the area are not locals, and thus do not appreciate the value of heritage they live in, and use their properties haphazardly.
- Limited open spaces to satisfy social and culture imperfections (Fig. 4.19).
- There is a shortage of ideologically relevant landmarks.
- Administration flaws on top of local authority efficiency shortages.
- Limited terminology of historical and cultural elements.
- Culture, heritage, identity, intimacy and history are in short supply.
- Little sense of pride, patriotism and belonging is raised through the built environment. The importance of named culture is hindered.

Fig. 4.19 Limited open spaces to improve social and cultural imperfections

- There is little to encourage public engagement and participation and mobilise the local assets.
- There is little to help strike the right balance between private and public spheres.
- Insufficient appropriate programmes to raise local awareness about merits of social interactions and engagement in the success of such projects.
- There is a shortage of considerations about human wellbeing and its role in improving people's socioeconomic status (Figs. 4.20 and 4.21).

Fig. 4.20 Using elements from culture to reflect the identity of the place

Fig. 4.21 Public participation and raising the awareness of residents

4.5 Conclusion

Research findings showed that in order to achieve user satisfaction (residents and visitors) in the historical district of Muharraq's, urban conservation projects, principles of LA have to be considered. The 'evaluation matrix' reveals that the absence of LA elements results in deficiencies in the overall performance of outdoor spaces in the pilot study. The matrix can contribute in spotting weaknesses of the design of LA elements that can be improved. This method can be used in similar projects and can result in provide clear action directions. Major amongst considerations of this method are time, movements, travel time, places for visitors, tourists and residents, within sustainable vision of such projects.

References

Abdel-Rahman O (1991) Visual notation in urban design, 1st edn. Liverpool Uni, UK, Liverpool

Al-Abdullah Mohammed M (1998) Relevance of the local people's socio-cultural values in the landscape development of recreational sea fronts of Saudi Arabia: the Case of "Dammam (1 ed). Un-Published Ph.D.. Landscape Architecture Thesis, University of New Castle Upon Tyne, UK

El Dardiry D (2013) Landscape architecture mechanism for sustainable urbanism. J Eng Sci 41 (3):1238–1258. May, 2013, 1551–1580, ISSN: 1687-0530 (print), ISSN: 2356-8550 (online). Available at: http://www.jes.aun.edu.eg/

Eldardiry D (2002) Urban development management in new towns in Egypt, 1st edn. Department of Architectural Engineering, College of Engineering, University of Menoffia, Menoffia, Egypt

El-Ghonaimy I (2005) Historical values and architectural unique building authentication. Arab City Magazine, Arab Towns Organization, Kuwait 126:50–66

El-Ghonaimy I (2009) Poverty and urban development effects of conflict in city downtown. J College Plann Cairo University, Egypt, 8

El-Ghonaimy I (2011a) Architectural heritage building between preservation and restoration (1 ed). Dar El-Koutab Publisher, Alexandria, Egypt

El-Ghonaimy I (2011b) Paradigm of landscape architecture realizing the sustainability in recreational urban areas. J Al-Azhar University Eng Sector (1)

Gareth Doherty CW (2016) Is landscape...? Essays on the identity of landscape, 1st edn. Routledge, New York

Geddes CL et al (1965) Studies in Islamic art and architecture in honour of professor. K.A.C. Creswell, Cairo, Cairo

Holden R, Liversedge J (2014) Construction for landscape architecture – book review. https://www.gardenvisit.com/blog/robert-holden-and-jamie-liversedge-construction-for-landscape-architecturebook-review/ ISBN: 9781856697088, published on May 4, 2011

Khan HH (2005) The Aga Khan Award official webpage. Accessed 24 April 2017, from www.AKDN.org

Maryland (1997) Association of historic district commissions (1 ed). Maryland

Murphy MD (2016) Landscape architecture theory: an ecological approach, 1st edn. Island Press, Washington

Planning TM (2005) Revitalization of Muharraq Historical Center. The Ministry of Municipalities Affairs and Urban Planning, Manama

White ET (1991) Space adjacency analysis, 1st edn. Architecture media Ltd., Flordia, USA

Part II
Governing Urban Heritage

Chapter 5
The Rise of the Facilitation Approach in Tackling Neighbourhood Decline in Tehran

Kaveh Hajialiakbari

Abstract One of the main urban problems of Tehran is the decline of its central neighbourhoods and the spatial concentration of deprivation within them. Since 2009, facilitation—encouraging open dialogue among stakeholders to explore diverse options—has been the main approach by the Urban Renewal Organisation of Tehran in tackling neighbourhood decline, to the extent that it is becoming the dominant public sector policy in Iran. In this way, more than 50 Facilitation Offices have been established within the declining neighbourhoods of Tehran and other cities. For this reason, it is necessary to document and analyse Tehran's experience, to recognise weaknesses and propose necessary readjustments to enhance the effectiveness of the approach. Based on this rationale, the main aim of this research is to document and analyse the main processes and outcomes of Facilitation Offices, and to identify the main weaknesses of the experience. Since the research is linked to the domain of urban policy, it follows the principles of policy research. Three major findings are of note: first, despite adopting an integrative perspective and involving the neighbourhood residents in the planning process, a lack of communication, coordination and consensus with non-local stakeholders is clear; second, the legal position of the neighbourhood regeneration plan is vague, with diverse agencies having no obligations to follow it; and third, in spite of the huge success of housing reconstruction projects, few examples of the provision of public services can be found.

Keywords Neighbourhood decline · Facilitation · Participatory planning · Land readjustment

---

K. Hajialiakbari (✉)
Shahid Beheshti University, SBU, Tehran, Iran
e-mail: kaveh_haa@yahoo.com

© Springer Nature Switzerland AG 2020
F. F. Arefian and S. H. I. Moeini (eds.), *Urban Heritage Along the Silk Roads*,
The Urban Book Series, https://doi.org/10.1007/978-3-030-22762-3_5

5.1 Introduction

5.1.1 Definition of the Problem

The problem can be defined here as the relative underperformance of many local urban economies and the resulting mix of economic, social, physical and environmental deprivation and exclusion. The result is that, without intervention, many urban areas appear to experience a self-sustaining downward spiral of decline in many respects (McCarthy 2007, p. 7). The conditions of poverty and exclusion interact and reinforce each other in particular geographical locations and make it impossible for families to escape negative 'neighbourhood' effects (Pierson 2002, p. 14).

In Iran, urban deterioration is defined as the inefficiency of a district in relation to other urban districts (Sharan Engineering Consultants 2005, p. 6). The appearance of deteriorated districts is explained based on three trajectories: first, the transformation of the country's economic base, rapid urban growth, the concentration of poverty in cities, and social exclusion; second, the failure of the real estate market and deprivation of the poor from affordable housing; and third, the failure of the public sector in economic and social policy-making, urban planning and local governance (Physical Development Institute 2010, p. 6).

In Tehran, a combination of the aforementioned reasons has led to the emergence of two major types of deteriorated neighbourhoods since the 1950s: first, the decline of the socio-economic heart of the city, the migration of primary inhabitants and the deterioration of the inner city; and second, the occupation of vacant lands around the city-centre by newcomers (especially the urban poor), the construction of non-standard houses, and the creation of new neighbourhoods without the necessary amenities (Hajialiakbari, forthcoming(a)). Neighbourhood deterioration in Tehran can be identified with certain symptoms: high population density,[1] high severity vulnerability against natural disasters (especially earthquakes), a lack of infrastructure and public services, narrow streets, increasing environmental pollution, high crime rate; and the low value of real estate (Sharan Engineering Consultants 2005, pp. 18–21).

---

[1] Population density; in deteriorated neighbourhoods in Tehran is 372 person per hectare and 2.9 times higher than the average of the city (127 person per hectare) (author based on (Statistical Center of Iran 2013)).

5 The Rise of the Facilitation Approach in Tackling Neighbourhood … 57

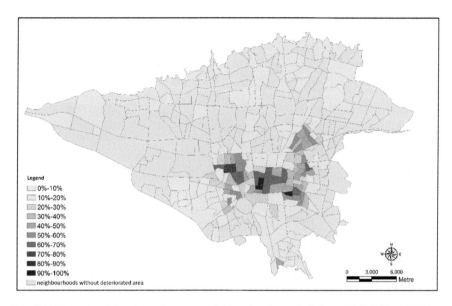

Fig. 5.1 The ratio of deteriorated areas to neighbourhood area in Tehran (Hajialiakbari 2017)

The recognition of deteriorated areas in Tehran[2] (in 2006) revealed that 196 neighbourhoods[3] have at least one deteriorated block.[4] However, 66%[5] of deteriorated areas[6] are concentrated in 56 neighbourhoods in the central parts of the city (Fig. 5.1) (Hajialiakbari 2017).

The importance of Neighbourhood Renewal has been prioritised by the municipality of Tehran since 2002,[7] and UROT[8] started to provide plans and implement projects in these areas between 2002 and 2009. Until the last decade, the main approach by the public sector in Iran was to intervene in urban districts authoritatively; this approach was based on top-down planning, taking coercive

---

[2]The indices of urban deterioration are defined by the Supreme Council of Architecture and Urban Planning of Iran in 2006 and are: vulnerability against earthquake; narrow streets (lower than 6 m); and small size of residential parcels (lower than 200 m^2). For more information about the indices of deterioration in Tehran, see: Boom Sazgan Engineering Consultants (2006).

[3]Tehran has 354 neighbourhoods.

[4]In official documents, a block is defined as an area restricted by roads from all sides. Boom Sazgan Engineering Consultants (2006).

[5]2,148 ha.

[6]3,268 ha of the area of Tehran are deteriorated. For more information about the scope of deterioration in Tehran, see: Boom Sazgan Engineering Consultants (2006).

[7]The report entitled 'The vulnerability of Tehran against earthquake' provided by JICA (the Japanese International Cooperation Agency) in 2002 was the first incentive for renewal in Tehran (Hajialiakbari 2011b, p. 26).

[8]Urban Renewal Organisation of Tehran was established in 1968 and is the main responsible agency for urban renewal in the municipality of Tehran.

Fig. 5.2 The construction of the edges of the Navvab Highway in Tehran (Andalib and Hajialiakbari 2008, p. 36)

possession of estates, banishing residents and implementing mega-projects. The construction of the edges of Navvab Highway in the 1990s was the most high-profile instance in Tehran (Fig. 5.2), with the huge scope, long implementation time, excessive financial resources and negative social impacts of such projects turned the approach into an unsuccessful, inefficient and unrepeatable experience (Hajialiakbari, forthcoming(a)).

Seeking new and efficient solutions to tackle neighbourhood deterioration, the Urban Renewal Organisation of Tehran (UROT) established a local office in a neighbourhood in 2007 and started to converse with the residents[9], with the main aim of the local office being to persuade occupants to participate in LRPs[10] (Hajialiakbari 2011a). After 2 years and the appearance of considerable outcomes from the experience, UROT decided to develop local offices in other neighbourhoods. This policy, named the 'facilitative approach', has led to the establishment of more than 50 offices in deteriorated neighbourhoods since 2009 (Hajialiakbari Forthcoming(b)).

---

[9]The Khoob-Bakht neighbourhood is one of the most deteriorated neighbourhoods of District 15 in Tehran. For more information about the experience, see: Andalib and Hajialiakbari (2008).

[10]Land Readjustment Projects consist of some adjacent parcels which are merged to provide enough land to construct a residential project. In Tehran, such projects have been implemented due to the small size of current parcels (60% less than 100 m^2) and the impossibility of receiving construction permits from the municipality (because of the regulations of the comprehensive plan of Tehran).

5 The Rise of the Facilitation Approach in Tackling Neighbourhood …

In 2009 and in tandem with the establishment of the first offices in Tehran, a legal procedure was announced,[11] which allowed municipalities to establish 'renewal service offices' in deteriorated and informal settlement neighbourhoods.[12] Since 2011, other municipalities,[13] based on this legal capacity and Tehran's experience, have started to establish new offices and the Urban Development and Rehabilitation Corporation (UDRC)[14] is planning to establish 100 Facilitation Offices in urban districts in Iran (UDRC 2016).

It can, therefore, be concluded that facilitation is becoming Iran's main approach for tackling urban deterioration, and so it is necessary to analyse the adopted process and achieved outcomes of Facilitation Offices. This analysis can help to document and analyse procedures, recognise weaknesses, enhance effectiveness and help responsible agencies utilise an improved framework.

5.1.2 Research Questions

Based on the above-mentioned rationale, the key questions of the research can be seen in Table 5.1.

Table 5.1 The key questions of the research

Type of question	Question
Descriptive	What are the main processes and outcomes of facilitation offices?
Analytic	What are the main weaknesses of the facilitation practice in Tehran?
Prescriptive	How can possible weaknesses be alleviated to raise the approach's efficiency in Tehran and other cities of Iran?

---

[11]'The rule of procedure of arrangement and support of housing provision and implementation' law was declared by the cabinet of Iran in 2009. In Article 42, municipalities were allowed to establish renewal service offices with the participation of private sector to accelerate renewal process. For more information about the law and its rule of procedure, see: The Cabinet of Iran proceedings (2009).

[12]The permission which was proposed by the Ministry of Housing and Urban Planning was based on past experiences of the ministry on urban renewal and establishment of neighbourhood offices in some projects such as Joolan neighbourhood renewal (Hamedan) and Joobare neighbourhood rehabilitation (Isfahan). For more information about these experiences, see (Aeini 2011) and (Civil and Housing Builders of Isfahan nd).

[13]Such as Isfahan, Shiraz, Tabriz and Rasht.

[14]The Urban Development and Rehabilitation Corporation is established in 1997 and is the responsible agency in Iran's government for urban regeneration.

5.1.3 Research Method

Since the research is linked to the urban policy domain (urban regeneration policy), it follows the principles of policy research. Policy research refers to the collection and analysis of information in order to inform the policy process (Maddison and Denniss 2009, p. 218), and is concerned with mapping alternative approaches and specifying potential differences in the intention, effect and cost of various programmes. The main features of policy research are as follows: the critical stance, the communicative process, a non-abstract product and consideration of malleability (Etzioni 1971, 2006).

Policy research uses diverse methods and techniques, such as quantitative methods (the process of inquiry into the quantitative elements of an issue or problem); qualitative methods (to provide a deeper understanding of behavioural motivation); and comparative methods (considering interstate, international and inter-temporal attempts to examine the relationship between two variables, or to examine the likely effectiveness of a range of policy options (Maddison and Denniss 2009, pp. 224–230; Maginn 2006, p. 2).

Hence, to gather and analyse required data, a combination of quantitative and qualitative techniques is used. In this study, a review of the planning documents and the proceeding reports of selected offices, and the preparation of in-depth interviews, FGDs,[15] and questionnaires with office staff, neighbourhood inhabitants and UROT directors are employed, and the content of the gathered data is analysed to compare the findings and propose necessary adjustments.

5.2 Definition of the Main Concepts of the Study

5.2.1 Neighbourhood Renewal

In 2000, the Social Exclusion Unit[16] presented the National Strategy for Neighbourhood Renewal. The aim of the strategy is to arrest the wholesale decline of deprived neighbourhoods, reverse it and prevent it from recurring. The two specific strategic goals are bridging the gap between the poorest neighbourhoods and the rest of the country, and lowering unemployment and crime, enhancing health and improving qualifications in all the poorest neighbourhoods (Social Exclusion Unit 2000, pp. 42, 44).

In Tehran and in 2007, the comprehensive plan of the city emphasised on encouraging policies to intervene in deteriorated neighbourhoods. The main policies

---

[15]Focus group discussions.

[16]The Social Exclusion Unit was set up by the Labour government in 1997 and formed part of the Office of the Deputy Prime Minister. The main responsibility of the unit was to provide the UK Government with strategic advice and policy analysis in its drive against social exclusion.

are: empowerment; the consolidation of neighbourhood organisations; integrative planning with consideration to environmental, social and cultural circumstances; land readjustment; provision of required services and infrastructure; and, the utilisation of diverse techniques such as rehabilitation, renovation and reconstruction (Boom Sazgan Engineering Consultants 2007, p. 11).

5.2.2 Participation in Neighbourhood Renewal Planning

As there are various actors involving in the regeneration process, there is a need to develop and to apply systems, procedures and techniques to facilitate collaborative and continuous management by diverse stakeholders. Therefore, in order to tackle the issues of urban regeneration, there are high expectations towards plan-making with the participation of various actors. Participation in the regeneration planning process means opening the contents of plan-making tasks to the public to make the plan accountable and transparent in terms of consensus building and decision-making processes (Murayama 2009, pp. 15–23).

In Iran, democracy and pluralism have been defined as the main principles of participatory planning in Neighbourhood Renewal. Participatory planning emphasises bottom-up processes, inclusion of all stakeholders and transferring necessary authority to neighbourhood agencies (Salehi and Vadoodi 2014, pp. 7–11).

5.2.3 Facilitation in Neighbourhood Renewal Process

Facilitation in community development is concerned with encouraging open dialogue among individuals with different perspectives so that diverse assumptions and options may be explored (Hogan 2002, pp. 10–11). Facilitation is about the process rather than the content and makes it easier to reach goals (Hunter et al. 1993, p. 5).

In Neighbourhood Renewal in Tehran, facilitation has been defined as the acceleration of the success of a group, the creation of interaction in the community, and provision of necessary structures to attain best outcomes through participation. In this context, the necessity for facilitation arises from the deprivation of local communities and the need for a mediator to begin and smooth the process of participation. Based on this definition, a variety of experts on social sciences, urban planning and design, architecture and economic sciences organise the structure of the Facilitation Office (Hajialiakbari et al. 2011, pp. 8–9).

5.3 Establishing Facilitation Offices in Tehran

5.3.1 Steps of Establishment and Expansion

One of the main problems of deteriorated neighbourhoods in Tehran is the small areas of current plots. Between 2005 and 2006, UROT started to provide renewal plans for some neighbourhoods based on merging and readjusting residential plots. One of these neighbourhoods was 'Khoob-Bakht' (Fig. 5.3). After the preparation of the renewal plan, a Neighbourhood Renewal office was established in 2007 to persuade residents to negotiate with each other to readjust their plots. The renewal office arranged more than 30 public and group meetings with the occupants and started the implementation of the first LRPs in October 2007. Until 2009, more than 200 owners participated in the programme and the implementation of 40 LRPs had begun (Andalib and Hajialiakbari 2008).[17]

In 2009, because of the considerable outcomes of the Khoob-Bakht renewal office, UROT decided to establish five new offices in some neighbourhoods which had renewal plans. The concept of 'facilitation' as the main approach of Neighbourhood Renewal offices was created at this stage and determined the future process of offices (Hajialiakbari et al. 2011). The expansion of Facilitation Offices continued over the next few years and the number of offices increased to 10 in 2010, 43 in 2011, and 63 in 2012; the activities of Facilitation Offices are still ongoing at the time of writing facilitation has become the main policy of UROT in tackling neighbourhood deterioration (Fig. 5.4) (Hajialiakbari, Forthcoming(b)).

5.3.2 The Process

The process of facilitation in Tehran is comprised of four phases (Fig. 5.5) (UROT 2015a). In the first phase—after UROT assesses, appoints and draws up a contract with the facilitator—the establishment of an office in the neighbourhood occurs. In this step, the office identifies the important stakeholders of the neighbourhood and communicates with them. The main aim of this phase is to introduce the office and its mission to the inhabitants and the neighbourhood organisations, form a dialogue, and build primary trust.

In the second phase, the neighbourhood's renewal plan is provided. The plan is prepared in three steps: first, recognition of the economic, social and physical characteristics of the neighbourhood using diverse methods such as observation, questionnaires, in-depth interviews and FGDs; second, the definition and prioritisation of the main problems and most important needs of residents; and third, the

---

[17]The main encouraging tools to persuade residents to participate in the plan were: free construction permit, low-interest loan for reconstruction, and zero interest loan for temporary settlement. For more information, see: (Andalib and Hajialiakbari 2008).

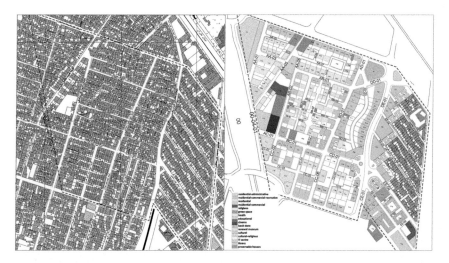

Fig. 5.3 Khoob-Bakht neighbourhood's status quo and renewal plan (Andalib and Hajialiakbari 2008, pp. 62–63)

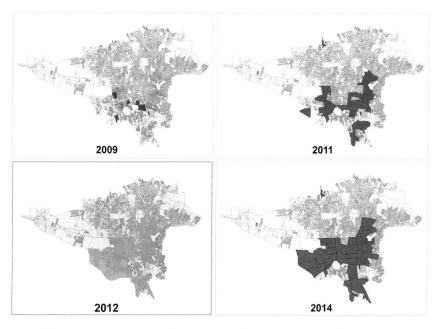

Fig. 5.4 The establishment of Facilitation Offices in deteriorated neighbourhoods of Tehran (Hajialiakbari, Forthcoming(b))

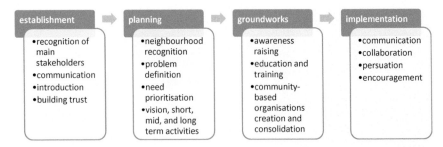

Fig. 5.5 The process of facilitation in neighbourhood offices in Tehran (Author, based on UROT 2015a)

suggestion of a plan, incorporating the neighbourhood's vision, short-term activities (up to one year), mid-term projects (between one and three years), and long-term programmes (between three and five years). It is expected that the plan is assessed and revised periodically (Hajialiakbari, Forthcoming(b)).

In the third phase, groundworks are led by the Facilitation Office. Groundworks are defined as a pre-condition for the participation of the local community and realisation of the neighbourhood plan (Hajialiakbari et al. 2011, p. 61), and can be divided into three fields (Hajialiakbari, Forthcoming(b)):

- Awareness-raising: without the agreement of residents about the inevitability of change, any intervention in the current circumstances of the neighbourhood cannot succeed. Therefore, Facilitation Offices utilise diverse tools to deliver their message to inhabitants. Some of these tools are as follows: public hearings, group and individual consultation meetings, a neighbourhood magazine, tracts, brochures and banners.
- Education and training: education can prepare residents for new circumstances and help them to activate their capabilities. Thus, Facilitation Offices organise educational courses about diverse subjects such as team working, living conditions in new apartments, women's role in renewal, and home-based jobs.
- Creation and consolidation of community-based organisations: it is assumed that Facilitation Offices' presence in the neighbourhood is temporal and—after creation and sustaining renewal flow—their responsibilities should devolve to community organisations; therefore, offices recognise and organise volunteers into distinct committees to communicate with other stakeholders and pursue anticipated projects.

In the fourth phase, the executive activities of the office commence. These activities are organised based on the Neighbourhood Renewal plan. Implementation of anticipated projects and programmes needs the participation, partnership and collaboration of inhabitants, private investors, public sector agencies and neighbourhood organisations. Hence, the office focuses on communication between, and satisfaction, persuasion and encouragement of, these stakeholders (Hajialiakbari, Forthcoming(b)).

5.3.3 The Outcome

The outcomes of activities can be divided into three fields: housing reconstruction, public services provision and public space rehabilitation (UROT 2015b).

5.3.3.1 Housing Reconstruction

Because of the threat of earthquakes, housing reconstruction has become the main policy of Iran in tackling urban deterioration since the 2000s. Low-interest loans for reconstruction and temporary habitation (Parliment of Iran 2008), grant free construction permissions (Tehran City Council 2009), and surplus density for new constructions in addition to the regulations of the comprehensive plan (Deputy of Urban Planning and Architecture-Tehran Municipality 2010) are the most important encouragement initiatives for land readjustment and housing reconstruction in deteriorated neighbourhoods in Tehran. Also, the small area of current residential plots (Fig. 5.6) have turned LRPs into the main agenda of Facilitation Offices (Hajialiakbari et al. 2010, pp. 14–15). Hence, these offices recognise capable projects and converse with occupants to persuade them to enter into partnership. They also assist in finding an investor to construct the project, prepare and conclude a partnership contract, and obtain a construction permit from the municipality. In some cases, offices also prepare the architectural design of the project.

Of the three types of possible LRPs (Fig. 5.7), small-scale projects are more popular, thanks to their fewer problems in persuading residents and their greater chances of realisation.[18]

An analysis of construction permissions shows that 1,184 LRP permissions were granted annually (on average) between 2009 and 2014 in deterioration neighbourhoods; this means that there was a higher than 2000% increase in LRPs compared with 2008 (Fig. 5.8).

5.3.3.2 Public Services Provision

Another problem in deteriorated neighbourhoods is their lack of public services (Table 5.2). Thus, Facilitation Offices spot essential needs based on the communities' demands. In the next step, offices negotiate with public sector agencies and persuade them to allocate a budget and implement necessary services (UROT 2015b).

---

[18]The average number of parcels in LRPs between 2009 and 2014 is 2.5 (UROT 2015c).

Fig. 5.6 General morphology of parcels in deteriorated neighbourhoods of Tehran (Hajialiakbari et al. 2010, p. 16)

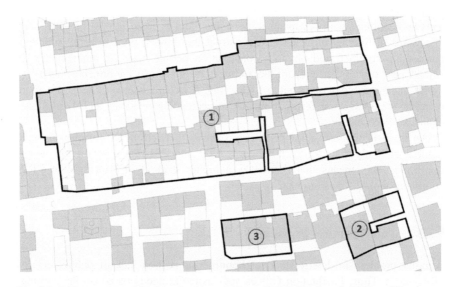

Fig. 5.7 Different types of land readjustment projects (Hajialiakbari et al. 2010, p. 30)

5.3.3.3 Public Space Rehabilitation

The absence of equipped urban space in the deteriorated neighbourhoods, and the importance of an attractive public domain as part of the renewal process, has led offices to plan the rehabilitation of current spaces in the Neighbourhood Renewal plan. Between 2009 and 2014, 50 rehabilitation projects were proposed to district municipalities and some proposals were implemented (such as improvements to

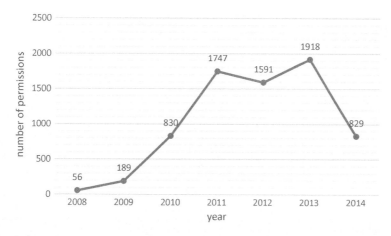

Fig. 5.8 The number of construction permissions of LRPs in deteriorated neighbourhoods of Tehran (Author, based on UROT 2015c)

Table 5.2 Area and per capita of land uses in deteriorated areas of Tehran

Land uses	Deteriorated neighbourhoods		Average of Tehran	
	Area (%)	Per capita (m^2)	Area (%)	Per capita (m^2)
Residential	62	16	29	23
Public services	4	1	8	6
Public paths	23	6	18	15

Source UROT (2006)

pavements and lighting on public paths, the reduction of unsafe spaces and the provision of neighbourhood parking) (Fig. 5.9) (UROT 2015b).

5.4 Analysis of the Experience

In this part, the process and outcomes of Facilitation Offices between 2009 and 2014 are analysed. Due to the number of offices (more than 50), ten neighbourhoods were selected as case studies (Fig. 5.10). The criteria for their selection included high, middle and low-grade offices,[19] the establishment of the office in neighbourhoods for a minimum of 2 years, an appropriate scattering of offices,[20] and diversity of neighbourhood characteristics.[21]

---

[19]Based on the assessment of UROT.
[20]In different neighbourhoods.
[21]Different social, economic and physical problems.

Fig. 5.9 Cases of implemented projects in urban spaces (UROT 2015b)

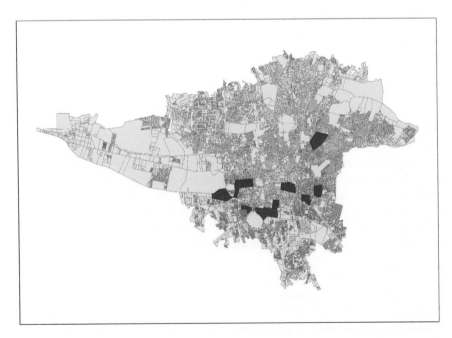

Fig. 5.10 Case studies analysed

5.4.1 Planning Documents

The analysis of planning documents and interviews[22] with planners in Facilitation Offices revealed some weaknesses in planning stages. The main weaknesses are:

- **The lack of consideration of the context**: The planning process and expected outcomes in all neighbourhoods are similar, and the singularities of each context and their neighbourhood characteristics and problems have a negligible impact on the planning document. This attitude has led to the provision of identical planning documents in all studied neighbourhoods.
- **The vague position of the planning document**: The Neighbourhood Renewal plan has no officially recognised position within the hierarchy of urban development plans in Tehran, and thus it is not sanctioned by legal authorities and cannot obligate diverse agencies to comply with plans' programmes.
- **The obliviousness of public agencies about the planning process**: Despite the participation of neighbourhood stakeholders in different stages of the planning process, public authorities (especially, district municipalities) have no distinct role in recognising problems, defining demands and approving programmes.
- **The lack of collaboration**: The realisation of the planning document needs the collaboration of different authorities in the public sector, but the compartmentalised approach of public agencies and lack of collaborative management in the renewal process hinder the achievement of some expected outcomes.
- **The shortage of financial resources**: The implementation of projects needs a public budget (the surplus of current expenditures of public agencies); however, insufficient financial resources are a serious obstacle to the execution of these projects.

5.4.2 LRPs

To analyse the role of the Facilitation Offices in LRPs, 200 projects in two groups—implementation with or without the mediation of offices—were selected and 389 questionnaires were filled in by the residents. The results of the analysis of questionnaires and the comparison of the two groups can be summarised as follows:

---

[22]Ten individual interviews and an FGD.

Table 5.3 Comparison of original residents in LRPs

Original residents?		Yes	No	Total
Projects of offices	Number	127	66	193
	Percentage	65.8	34.2	100
Other projects	Number	89	81	170
	Percentage	52.4	47.6	100

Table 5.4 Comparison of the area and number of parcels in LRPs

Specification of projects		Average
Area of projects (m^2)	Projects of offices	217.8
	Other projects	186.7
Number of parcels in projects	Projects of offices	3.4
	Other projects	2.6

The sustained habitation after reconstruction: The most important factor in housing reconstruction projects is the sustained habitation of the current occupants. The analysis of this factor shows that in the Office projects, more apartments kept their original inhabitants (Table 5.3).[23]

The area and the number of readjusted plots: A comparison of the number and area of merged plots between those in projects realised through Offices and the rest shows that these figures were higher in the former by 17 and 30%, respectively (Table 5.4). This means that Facilitation Offices perform well in persuading residents into partnership arrangements.

The satisfaction with the reconstruction: A comparison between satisfaction levels with living circumstances in new apartments and old houses) shows that the levels improved in both groups of projects (Table 5.5),[24] but the increase in the Office projects was sharper.

The quality of social relationships: The comparison of social relationships with neighbours shows that readjustment reduced relationships, and there was no difference between the projects of Facilitation Offices and other projects (Table 5.6).

The quality of construction: The analysis of partnership contracts drawn up by Facilitation Offices shows that in these contracts (in comparison with regular contracts) a better quality of technical specifications was considered and investors were obligated to construct projects with the best possible materials.

---

[23] Financing and construction of LRPs in the deteriorated neighbourhoods of Tehran is done by a private investor, so, due to the value of land, between 40 and 50% of constructed apartments belong to the investor.

[24] Attaining a '1' in Table 5.5 means a better level of satisfaction.

5 The Rise of the Facilitation Approach in Tackling Neighbourhood … 71

Table 5.5 Comparison of the level of satisfaction from living conditions in LRPs

Projects	Before readjustment	After readjustment
Projects of offices	1.78	1.55
Other projects	1.24	1.21

Table 5.6 The quality of social relationships before and after readjustment

Social relationship with neighbours	Before readjustment (%)	After readjustment (%)
Very good	24	15.9
Good	55.2	49.6
Mediocre	20.4	31
Bad	0.5	3.5

5.4.3 Public Services

The analysis of the proceeding reports of selected Facilitation Offices shows that, despite the anticipation of required services in planning documents, few cases of amenities provision have been realised. Due to the increase in the number of residential units, the population of neighbourhoods also increased after renewal, and this situation can decrease the accessibility of public services in per capita.

A comparison of the population and per capita of public services in selected neighbourhoods shows that if all residential plots are reconstructed, the population of neighbourhoods will increase by 65%: under these circumstances (the non-provision of necessary services), the per capita of all public services will decrease (Table 5.7) (Hajialiakbari, Forthcoming(b)).

Table 5.7 Comparison of population and per capita of public services in current circumstances and after renewal

Factor	Current circumstances	After reconstruction
Population of neighbourhoods (person)	232,936	385,966
Parks (per capita)	0.27	0.16
Sporting land uses (per capita)	0.08	0.05
Cultural land uses (per capita)	0.28	0.17
Educational land uses (per capita)	0.61	0.37
Health land uses (per capita)	0.09	0.05

5.4.4 Urban Space Rehabilitation

The analysis of planning documents and proceeding reports of selected Offices shows that 25 rehabilitating projects were anticipated between 2009 and 2014, and except for one case,[25] other projects were not implemented (Hajialiakbari, Forthcoming(b)). The officially unrecognised position of planning documents and thereby the lack of obligation in district municipalities to implement these documents were behind these circumstances.

5.5 Conclusion

The establishment of Facilitation Offices in deteriorated neighbourhoods of Tehran is a progressive approach which utilises participatory and bottom-up planning methods to involve stakeholders in the renewal process. The main outcomes of this approach are the prevention of coercive possession and banishment of residents, a planning process at the neighbourhood scale, a more comprehensive approach (especially with regard to social aspects) to Neighbourhood Renewal, the devolution of the reconstruction, the consideration of necessary public services, and emphasising the importance of community-based organisations in the renewal process.

The analysis of the experience also reveals weaknesses which should trigger a revision of certain components of the approach. The main suggestions of this research for improving the efficiency and effectiveness of the approach are as follows:

- **Contextualisation**: Facilitation cannot ignore specific characteristics and problems of each neighbourhood and its residents. Thus, the outputs of planning depend on the uniqueness of context. For example, the analysis of LRPs shows that the quality of social relationships decreases after reconstruction, and so the definition of housing projects in a neighbourhood must consider cultural and social values and the principles of residents.
- **Inclusiveness**: The presence of stakeholders in the planning process is not restricted to residents and neighbourhood organisations. All actors with a role in the renewal process must be included (such as public agencies and non-local stakeholders).
- **Legitimisation**: The Neighbourhood Renewal plan cannot be realised without maintaining a legal position in urban planning hierarchy and without the obligating all stakeholders to participate in implementation of the plan.

---

[25]The 'Green Neighbourhood' project in District 14.

- **Equilibrium**: The renewal of a neighbourhood includes diverse dimensions which must be led simultaneously; the emphasis on a dimension (such as housing reconstruction) and the negligence of other dimensions (such as public service provision) can decrease living standards and accelerate deprivation in deprived neighbourhoods.
- **Collaboration**: The renewal process has an interdisciplinary nature: the provision of public services, reconstruction of houses, establishment of community-based organisation, and rehabilitation of urban space that settle Neighbourhood Renewal in the intersection of local communities and public, private and third parties. As a result, the collaboration of these parties in all steps of the process is unavoidable.
- **Financing:** The analysis of the previous experiences suggests that, in spite of a considerable increase in housing reconstruction projects (run by the partnership of residents and private investors), anticipated programmes in the public domain have not been realised. Thus, the realisation of a Neighbourhood Renewal plan needs the preparation and allocation of the required budget by public sector authorities.

References

Aeini M (2011) The experiment of neighbourhood renewal office in Joolan. Nosazi Online
Andalib A, Hajialiakbari K (2008) Renovation of deteriorated areas with the participation of residents. ROT, Tehran
Boom Sazgan Engineering Consultants (2007) The comprehensive plan of Tehran. The supreme council of architecture and urban planning of Iran, Tehran
Civil and Housing Builders of Isfahan (nd) Rehabilitation of Joobare neighbourhood. Accessed June 03, 1394, from Civil and Housing Builders of Isfahan. http://www.maskansazancz.ir/Default.aspx?tabid=108
Consultants Boom Sazgan Engineering (2006) Deteriorated areas of Tehran. Supreme Council of Architecture and Urban Planning of Iran, Tehran
Deputy of Urban Planning and Architecture—Tehran Municipality (2010) Regulations of detailed plan of Tehran. Municipality of Tehran, Tehran
Etzioni A (1971) Policy research. The American Sociologist, pp 8–12
Etzioni A (2006) The unique methodology of policy research. In: Moran M, Rein M, Goodin RE (eds) The Oxford handbook of public policy. Oxford University Press, Oxford, pp 833–843
Hajialiakbari K (2011a) A review on the ROT's experiment in local offices of renovation. Nosazi Online
Hajialiakbari K (2011b) Facilitation in the deteriorated areas of Tehran. Haft Shahr, pp 24–38
Hajialiakbari K (2017) Definition of neighbourhood development framework in deteriorated neighbourhoods of Tehran. Tehran Urban Research and Planning Center, Tehran
Hajialiakbari K (Forthcoming (c)) Defining effective indexes of functional sustainability of deteriorated neighbourhoods in Tehran. Bagh-e Nazar
Hajialiakbari K (Forthcoming (a)) Facilitation in deteriorated areas of Tehran. In: Shirazi M, Falahat S (eds) Participation in Iran. Routledge, London
Hajialiakbari K (Forthcoming (b)) Facilitation in Tehran. UDRC, Tehran
Hajialiakbari K, Fallahzadegan M, Asgari H (2010) Readjustment of small-area parcels in deteriorated neighbourhoods of Tehran. Urban Renewal Organisation of Tehran, Tehran

Hajialiakbari K, Fallahzadegan M, Asgari H, Ghavampoor E, Laylavi F (2011) Facilitation; establishment of renewal facilitation offices in deteriorated areas of Tehran. Urban Renewal Organisation of Tehran, Tehran

Hogan C (2002) Understanding facilitation; theory and principles, 1st edn. Kogan Page, London

Hunter D, Bailey A, Taylor B (1993) The art of facilitation. Tandem Press, Auckland

Maddison S, Denniss R (2009) An introduction to Australian urban policy: theory and practice. Cambridge University Press, Cambridge

Maginn PJ (2006) Urban policy analysis through a qualitative lens: overview to special issue. Urban Policy Res 1–15

McCarthy J (2007) Partnership, collaborative planning and urban regeneration. Ashgate, Hampshire

Murayama A (2009) Toward the development of plan-making methodology for urban regeneration. In: Horita M, Koizumi H (eds) Innovations in collaborative urban regeneration. Springer, Tokyo, pp 15–29

Parliment of Iran (2008) The law of 'arrangement and support of house construction and supply. Research Center of the Parliament of Iran, Tehran

Physical Development Institute (2010) Definition of deterioration causes in Iran. Urban Development and Rehabilitation Corporation, Tehran

Pierson J (2002) Tackling social exclusion. Routledge, London

Salehi A, Vadoodi S (2014) Planning; education of strategic and executive planning to renovation service offices. City Publication Institute, Tehran

Sharan Engineering Consultants (2005) The manual of recognition and intervention in deteriorated areas in Iran. Iran supreme council of architecture and urban planning, Tehran

Social Exclusion Unit (2000) A national strategy for neighbourhood renewal; a framework for conclusion. Crown, London

Statistical Center of Iran (2013) National census of 2011. Statistical Center of Iran, Tehran, Tehran, Iran

Tehran City Council (2009) Financial encouragement of owners and investors in readjustment and renovation of deteriorated areas in Tehran. Tehran City Council, Tehran

The Cabinet of Iran (2009) The rule of procedure of 'arrangement and support of housing provision and implementation' law. Islamic Parliament Research Center of IRAN, Tehran

UDRC (2016) The annual program of neighbourhood regeneration in Iran in 2016. Urban Development and Rehabilitation Corporation, Tehran

UROT (2006) Analysis of the local services in the deteriorated neighbourhoods of Tehran. Urban Renewal Organisation of Tehran, Tehran

UROT (2015a) Contracts of the facilitation offices on 2009–2015. Urban Renewal Organisation of Tehran, Tehran

UROT (2015b) The annual reports of activities of facilitation offices. Urban Renewal Organisation of Tehran, Tehran

UROT (2015c) The portal of construction permits in deteriorated areas of Tehran. Urban Renewal Organisation of Tehran, Tehran

Chapter 6
Rebuilding Tajeel: Strategies to Reverse the Deterioration of Cultural Heritage and Loss of Identity of the Historic Quarters of Erbil, Kurdistan, Iraq

Anna Soave and Bozhan Hawizy

Abstract The inclusion of the Erbil Citadel in the Tentative List of World Heritage Sites in 2010 prompted the issuance of urban design guidelines aimed to ensure that buildings located within the Citadel's Buffer Zone respect the visual integrity of the citadel and its relationship with its setting. Concerned that it would have not been possible to effectively control the destruction of historic houses for more profitable commercial buildings, in 2013 local authorities expropriated all residential properties within the historic quarters. Alas, without a solid implementation plan and funding strategy in place, what might have been partially lost to real estate speculation is now inexorably collapsing under the rain. In the past decade, Erbil has seen the implementation of several ambitious urban renewal efforts. Community heritage has been pulled apart to create a new 'brand' for Erbil. The current economic downturn has laid bare the ambition to create a revamped image of Erbil that would attract domestic and foreign investments. Kurdish decision-makers are now struggling to rehabilitate urban heritage with very little public funding, convince citizens and developers to invest in these now empty and decaying quarters. The chapter explores how successful urban projects do not have to rely solely on government funding or large-scale developers. The proposed integrated area management plan takes into consideration the area's socio-economic potential and its legal, financial and environmental constraints. It suggests a set of pragmatic funding and implementation strategies for the rehabilitation of key historic buildings and infrastructure upgrading through cross-subsidisation and leveraging land value. It also advocates for a set of integrated measures to stimulate the economy by promoting small-scale commerce, enterprises and hospitality activities, generate municipal revenue, enhance the quality of life and finally nurture cultural and creative industries.

A. Soave (✉) · B. Hawizy
UN-Habitat Iraq Programme, Ebril, Iraq
e-mail: anna.soave@un.org

B. Hawizy
e-mail: bozhan.hawizy@un.org

Keywords Expropriation · Historic quarters · Rebuilding · Urban renewal historic cities · Iraq · Kurdistan · Erbil · Tajeel

6.1 Introduction

The inclusion of the iconic Erbil Citadel built on a natural mound located at the heart of the capital of Kurdistan, northern Iraq, in the Tentative List of World Heritage Sites in January 2010, prompted the local authorities to issue in partnership with UNESCO a set of urban design guidelines that aim to ensure that the buildings located within the Citadel's Buffer Zone respect the visual integrity of the citadel and its relationship with its setting. Concerned that it would have not been possible to effectively control the piecemeal demolition of dilapidated historic houses in favour of new more profitable multistorey commercial buildings, in 2013 the local authorities proceeded to expropriate all of the historic residential properties within the historic quarters. Alas, without a solid implementation plan and funding strategy in place, the rehabilitation of these fragile areas was put on hold. Regrettably, with no residents undertaking regular maintenance of their homes, what might have been partially lost to real estate speculation is now inexorably collapsing under the rain (Fig. 6.1).

In the past decade, Erbil has seen the implementation of several ambitious urban renewal efforts in the city centre. Community heritage has been pulled apart to create a new 'brand' for Erbil, to appeal to visitors and tourists. The current economic downturn has laid bare the ambition to create a revamped image of the core of Erbil that would attract domestic and foreign investments. Kurdish decision-makers are now faced with the major challenge of finding a way to rehabilitate its urban heritage with very little public funding, convince citizens and developers to invest in collapsed property and bring back life to these now empty and decaying quarters.

Fig. 6.1 Without its inhabitants the overall building stock and the existing infrastructure are rapidly deteriorating

This chapter seeks to explore how Kurdish planning authorities could implement new innovative urban projects that do not have to rely solely on government funding or large-scale developers. It builds upon consultations and participatory planning workshops initiated in mid-2015 by the planning staff of the Ministry of Municipality and Tourism (MOMT), Erbil Municipality and the Erbil Directorate of Urban Planning, working under the guidance of the authors under the umbrella of the 'Strengthening Urban and Regional Planning in KR-I' (SURP) project, implemented by UN-Habitat Iraq.[1]

6.2 Historical Background and Urban Context

6.2.1 The Erbil Citadel

The most distinctive landmark of Erbil is its Citadel, also known as the *Qala*, located at the very centre of the concentric ring roads for which the Kurdish capital city is widely renowned. The 10-ha fortified elliptical settlement sits on top of an earthen *tell* (artificial mound). Its origins, although uncertain, can be traced back to at least 6,000 years ago (Al-Hashimi and Bandyopadhyay 2015, p. 49). Within its walls, the houses have been rebuilt over the centuries, yet the narrow street pattern, *cul-de-sac* alleyways and its three functional *mahallas* (districts) have largely survived. Some 500 traditional courtyard houses are still standing, only a few of which in a good state of conservation, including some fine examples of residential buildings dating back to the nineteenth–twentieth century, and a few from the 18th century (ICOMOS 2010).

From the 1930s onwards, the Qala has witnessed the ongoing abandonment of its original inhabitants. As standards of living changed, services and infrastructure networks began to deteriorate; families began to leave the citadel to settle in the

---

[1]The 2-year SURP project, funded by the Kurdistan Regional Government (KRG), was designed to strengthen the technical and institutional capacity of the urban planning staff of MOMT and the Directorate of Urban Planning of the Erbil Governorate in view of the anticipated challenges of rapid urbanisation of this region. SURP included a 'learning by doing' training module that focused on the rehabilitation of a dilapidated historic area. Trainees were mentored through rapid assessments, urban diagnostics of the physical, legal, administrative and financial challenges that have hampered any rehabilitation action to date, an assessment of building conditions and status of occupancy, leading to the development of an integrated proposal. Widening consultation workshops involved colleagues from the Erbil Urban Planning Directorate and the Buffer Zone Committee of Municipality One, practitioners from UNESCO and the High Commission for the Erbil Citadel Revitalization (HCECR). A final report, titled *Integrated Rehabilitation and Management Proposal for the Historic Quarters of Tajeel*, was submitted in late 2016. The authors wish to acknowledge the contributions of the staff of MOMT, led by Mr. Abdulmomin Maroof General Director of Urban Planning, as well as colleagues of Erbil Municipality 1.

plain. In the late 1950s, the southern Grand Gate was demolished and a wide vehicular axis was cut through the dense urban fabric of the Qala from north to south. Many historic buildings were destroyed in the process. The southern gate was redesigned and built in 1979.

By the second half of the twentieth century, many poor tenants moved in and several empty homes were squatted. From 1986 the Qala served also as a refuge for families displaced by the genocidal Anfal campaign. By 2006, all inhabitants were relocated to allow for the restoration works to begin, save for one family, who was asked to stay in order to maintain the Citadel's record of continuous habitation (Michelmore 2013, p. 8). Despite the ongoing works and partial cordoning off of unsafe areas, the Qala remains a popular venue for gatherings, Friday prayers, visits and cultural events.

6.2.2 The Qaysary Bazaar

Branching off from the Grand Gate of the Citadel, at the foot of the *tell* lies the **Qaysary Bazaar**, dating back to the twelfth century. The convergence of routes towards a focal point other than a mosque is particular to Erbil (Al-Hashimi and Bandyopadhyay 2015)—otherwise, its urban structure is typical of an Islamic *suq*, with its covered *qaysari* (arcades) linear development of shops on the ground floor and accommodation and storage on the upper floor. The later demolition of the *hammam* and the *khans* along the main roads and the introduction of modern structures and car parks brought a rupture and change to the *suq*'s fabric.

6.2.3 The Residential Quarters of Arab, Tajeel and Khanaqa

At the foot of the citadel, fanning out from the *Qaysari bazaar,* lies the historic residential district comprising of the *Arab, Tajeel* and *Khanaqa* quarters. Reportedly, these settlements are at least 200 years old. Unfortunately, much of the information available is anecdotal. While the 'Arab' quarter took its name from the Arab immigrants who settled in this area, 'Tajeel' was inhabited by 'Jewish' and 'Kurdish' families who were said to cohabit peacefully.

With nearly 5,000 people recorded in the 1947 census, the population of Tajeel increased threefold in 1984 and then declined in 1997 to a little over 10,000. The area was a working-class neighbourhood, well known for its Jewish traders, goldsmith, carpet weavers, tailors, craftsmen, including architects and builders. Traces of the distinctive decorative plaster and brick craftsmanship of the time are still visible in the area. In its early days, according to the *mukhtar* (area representative), everybody knew each other since most of its inhabitants had been living there since the beginning. But as newcomers from surrounding villages, as well as Syrians, Iranians and Iraqi Arabs, came to live and work in Tajeel the social fabric started to change.

The urban fabric of the historic quarters is compact, served by an irregular and narrow street pattern. Spaces offer a range of public, semi-public, semi-private and private spaces, typical of the Middle East. Historically, social interactions and relations shaped the urban identity and sense of intimacy of these closely-knit quarters. Buildings were mostly flat-roofed, one or two-storey high. Spaces were typically organised around a central courtyard (*al-haush*) with porticos and awnings protecting internal spaces from the rain in winter and the heat in summer. The traditional construction material consisted of dried mud bricks, occasionally arranged in geometric patterns above doors or on upper parapets. The exterior of houses was rather plain, dominated by one entrance and few windows (Khasro et al. 2016, p. 134). The elevations often include shading elements, *shanasheel* balconies (a timber lattice feature typical of Iraq), *sitara* (parapet) and recesses corresponding with doors. In the 1960s and 70s, many traditional houses were rebuilt in concrete but largely maintained the courtyard typology.

6.3 Conservation and Regeneration Efforts

6.3.1 The Nomination of the Citadel as World Heritage Site

In recognition of the uniqueness and outstanding universal value of its urban heritage, the Erbil Citadel was officially inscribed as 'Archaeological Heritage of Iraq' in 1937.[2] Unfortunately, legislation did not prevent physical degradation, nor the intentional demolitions that occurred in the late 1950s (HCECR 2014, p. ii). Conservation works only started in 1978, yet there were few concrete initiatives until 2004 (Ibid). In 2007, Kurdish Regional Government (KRG) established the High Commission for Erbil Citadel Revitalization (HCECR), as the authority in charge of its revitalization efforts. Since then, HCECR and UNESCO have been closely collaborating to preserve this important site.

The *Conservation and Rehabilitation Master Plan* drafted in 2011 embraces the formula of 'adaptive reuse' relying upon a mixed-use approach based on cultural activities, tourism-related businesses, recreation and residential areas (Ibid., p. 10). Its 2012 *Management Plan* defines policies, strategies and actions aimed to restore

---

[2]The first protection Law on ancient monuments in Iraq dates back to 1936 (No. 59).

in phases the role and position the Citadel as '*a living place central to the life of the city of Erbil and the northern regions of Iraq, and as an urban landscape of importance for all humanity*' (UNESCO). In June 2014, despite the recognition that its building stock has suffered from serious discontinuity of its physical, social[3] and functional integrity due to several changes to its urban structure—including a new axis built in the twentieth century—the Citadel was finally nominated '*World Heritage Site*' by UNESCO (HCECR 2014).

6.3.2 The Creation of the Erbil Citadel Buffer Zone

By the time of the inclusion in the Tentative List of World Heritage Sites, UNESCO and HCECR had already successfully managed to convince the Kurdish government to adopt a substantially different vision from that offered by a Beirut-based consultancy company[4] in their 2006 Erbil Master Plan which proposed a radical transformation of Erbil downtown with modern avenues and elegantly clad buildings.[5] Besides, one of the key requirements of the UNESCO application was to establish a Buffer Zone around the Citadel with clear guidelines and regulations for the historical quarters, which by that time had suffered problems of integrity due to modern and incongruous constructions (ICOMOS 2010).

Local authorities worked hard to meet the conditions of the nomination. With the support of UNESCO, an international consultancy firm[6] was tasked to draft a set of *Urban Design Guidelines for the Buffer Zone of Erbil Citadel* to ensure that building development within this area reinforces 'the centrality of the ErbilCitadel as a key cultural and visual reference' and to promote the integration between the Citadel and its surroundings 'by controlling development along main radial roads with views to and from the Citadel, and by encouraging small strategic interventions to introduce tourist and cultural uses in the historic districts at the foot of the Citadel mound' (HCECR 2014, p. 156) The Guidelines, adopted in 2011, offer strict height restrictions intended to protect the visual corridors towards the citadel and detailed regulations aimed at preserving and enhancing the character of this area.[7]

---

[3]As per the Citadel's Nomination report, the relocation of all its inhabitants has 'unfavourably affected the social and functional integrity of the urban fabric as a traditional organically-evolved urban settlement'. ICOMOS (2010, p. 82).

[4]Dar al-Handasa, Lebanon.

[5]Although the buffer zone does not fall under the responsibility of either HCECR nor the Erbil Governorate (who is funding the conservation works of the Qala), but of Erbil Municipality, all institutions have demonstrated a close interest in the future of the historic areas.

[6]ARS Progetti, Italy.

[7]The Guidelines were revised in 2013 to offer even more control over construction heights and character.

6.4 Recent Urban Transformations that Affected the Historic Quarters

6.4.1 Erbil City Centre Restyled as a Tourist and Shopping Destination

Aside from the rehabilitation of the Citadel, in the last decade, the city centre has been the focus of several government-led urban regeneration efforts. In 2009 the informal *Delal Khaneh bazaar* was demolished to make spaces for the new *Shar Square*. The merchants, who reportedly resisted the project, were offered different relocation options either in the outskirts of the city or in the nearby Nishtiman bazaar, where some ended up renting spaces in the underground parking—with a significant loss of customers, rupture of social networks and increased rents. The new *Shar Square* was built in 2010 and has since then become a very popular meeting place for citizens and visitors alike. Surrounded by cafés and tea shops under the arcades, it is embellished by fountains, plants and benches for people to sit and enjoy the spectacular view of the Citadel walls and Grand Gate.

To the west, the old *Qaysary Suq* was given a radical facelift by the municipality in 2012 as part of its effort to harmonise what were perceived as incongruent frontages. The bazaar was clad in uniform ochre bricks, supposedly reminiscent of the local traditional style. New paving, water, lighting and roofing were introduced, greatly improving its internal spaces.

Just across the square, to the south, the multistorey *Nishtiman Bazaar* was built. The upper floors of the commercial complex have remained vacant for years, partly because of the reluctance of people to shop on upper floors. The building cordons off the most significant *cemetery* of the city centre, known as the *chragh*, part of which was removed to make space for future development (Fig. 6.2).

Framing the southern section of the cemetery, beyond the Nishtiman Bazaar, in 2013 the UAE-based Emaar real estate development company launched the *Downtown Erbil Shopping Centre*, a 540,000 m^2 mixed-use development reportedly worth USD 3 billion. The construction of the curved central axis, shops on multiple floors, internal parking and bus terminal was completed in mid-2016, offering the city centre an impressive new commercial destination. For the time being, the proposed glass-clad skyscrapers that were designed to host 100,000 m^2 of upmarket office and hotel space remain on paper, grounded by the height restrictions enforced on the Citadel Buffer Zone. Despite the developer's very influential connections, the scheme may have to capitulate on the construction of the towers and hopefully also improve the physical integration between the mall and the busy Shekalla bazaar which was 'walled off' in a blatant effort to restrict access from what is perceived as a 'lower income' area.

As for the nearby *residential historic quarters*, these were to be transformed in a prime cultural and touristic destination by means of a large real estate operation. Reportedly, concerned that it would have not been possible to effectively control the demolition of the historic fabric and the construction of inappropriate and

Fig. 6.2 New Shar Square built in 2010, photo of 2015

out-of-scale commercial buildings, the local authorities proceeded to expropriate all properties in 2013. Expropriation for the purposes of public benefit in Iraq is regulated by Acquisition Law No. 12 of 1981. The law aims to determine fair compensation for all property acquisition and guarantees the rights of possessor without prejudice to public interest. It foresees the possibility to offer the expropriated owner an alternative property or other real estates as compensation for the property requested for acquisition. Yet, there was neither a responsive implementation plan nor a funding strategy for the rehabilitation. With the loss of its precious socio-economic base, the overall building stock has rapidly deteriorated. Today, most of the houses are vacant—save for a few pockets where residents have resisted relocation.

6.4.2 Expropriation—The Ultimate Building Control Tool?

Both UNESCO and ARS Progetti had strongly opposed expropriation (only to be used as a 'last resort'), but this did not deter authorities from forcing residents out of their homes in Tajeel, Arab and Khanaqa. The KRG allocated each displaced family a plot of land 10 km outside of Erbil, along with a $4,000 contribution to build a new house (Monk and Herscher 2015, p. 68).

The motives leading to the government's decision to expropriate are not devoid of controversy. Whereas at the time of the expropriation of the Citadel properties, most original families had already moved out of their homes for many decades, attracted by better living conditions elsewhere, with the replacing people having no social or cultural roots there, the circumstances of the historical quarters located in the Buffer Zone were quite different. They were fully inhabited—aside from the fact that the building stock was of lesser historical and architectural value.

Interviewed by a researcher about the impact of gentrification in Erbil, the Governor justified the expropriation by affirming that property owners 'might have built commercial and high-rise buildings, which would have destroyed the heritage value of the area'. Questioned about the expropriation, the *mukhtar* said that residents largely opposed it, not only because people did not want to leave their properties, but also because they did not agree with the land compensation, which was deemed unfair in terms of valuation and the loss of basic services (Khoshnaw 2016).[8] Reportedly, residents would have preferred to stay and renovate their homes, but that option was not on the table. '*It was all our fault—stated a respondent—because when the government started the upgrading works, many families of the area blocked the labourers*'. According to another respondent '*those people were mostly tenants who were afraid of being excluded from the compensations scheme*'.[9]

Despite the strong disapproval towards the expropriation order, it was met with general resignation.[10] According to the *mukhtar*, today there are still some 174 inhabited houses, 32 of which owner-occupied. A rapid survey confirmed the survival of pockets of residents resisting relocation and the presence of newcomers (mostly poor displaced families seeking accommodation in condemned buildings). Fearing that the area would soon become a slum ridden by criminals and drug addicts, residents began to build concrete block walls to prevent break-ins.

A ban was imposed against any renovation and sale of homes, but reportedly ignored in some cases. While the expropriations are still being rolled out, earlier cases have not been yet cleared by the courts. According to the *mukhtar* and the elders interviewed, property owners would agree to renovate their homes according to their traditional typology. Regrettably, what might have been partially lost to real estate speculation is now inexorably collapsing under the rain.

---

[8]Interview conducted in February 2015.

[9]Finally, also the tenants managed to receive a small land in Shamamik, outside Erbil, where some built a home, but others sold it and sought rental accommodation in the city centre again.

[10]The team attempted to map the property expropriation cases initiated by the Government in 2012 to draft a Land Ownership Map—essential for any further strategic discussion on implementation mechanisms and land transfer—but the information was impossible to attain from the Governorate due to the lack of a database or updated maps.

6.4.3 Makeover Efforts—In Whose Interest?

Historic cities inevitably transform themselves. While it is typically an organic process of change over time, in Erbil such urban renewal efforts have often implied the displacement of many residential and small commercial activities from within the city centre. Community heritage has been pulled apart to create a new '*brand*' for Erbil, in an effort to appeal to national visitors and foreign tourists. Driven by what some researchers have perceived as a '*pervasive neoliberal discourse*', the redistribution of urban spaces appears to have been influenced by the socio-economic status of the users or residents—'*a new spatial fix where the urban poor are purged from the city centres and replaced with privatised and modernised urban projects that are considered a spatial cure for the concentration of the poor*' (Mohammadi 2014, pp. 2–4). Figures 6.3 and 6.4 present the example of the new Erbil Downtown development.

The harmful social and cultural repercussions of insensitive urban renewal projects that have implied forced relocations in other international cases is well understood, for example, in Istanbul, Turkey, where urban renewal and modernisation efforts have promoted the notion of an 'hygienised' city. In the Iraqi Kurdistan it seems that the unregulated *bazaaris* and the poor have become the new undesirables and are being squeezed out by the commodification of public land

Fig. 6.3 Finishing works in the Plaza fronting the new Erbil Downtown development

Fig. 6.4 Interior of the new Erbil Downtown development

(Oz 2012, pp. 297–314) and government-led gentrification of spaces. In the light of this trend of social exclusion applied to spaces, after the removal of the Delal Khaneh bazaar, the next place that risks being 'purged' from the city centre, is the colourful open-air fruit and vegetable market of Shekalla (Fig. 6.5).

While expropriation is grounded in an understanding that private property rights must sometimes be ceded for projects of 'public interest', expropriation in Erbil appears to have been applied more often to poor and middle-class neighbourhoods than to wealthier ones. International experience, such as the case of historical Beijing (Foster 2010), shows that when such decisions are being taken without an appropriate dialogue with people, expropriation risks being perceived to have been misused to benefit real estate developers who are conveniently exonerated from paying the true cost of land because of the perceived trickle-down 'benefits' of their

Fig. 6.5 The vegetable market of Shekalla attracts shoppers from all Erbil (Photo of 2015)

investments—as it clearly emerged from conversations held with the management team in charge of the new Erbil Downtown development. In the case of Tajeel, should the authorities go ahead with the sale of development rights, they could face a moral predicament, accused of seizing land from citizens in the interest of developers or wealthy investors.

The desire to create a revamped image of Erbil that would benefit both national and international tourism and attract the much needed domestic and foreign investment in the past 10 years, appears to have been the origin of the extensive acquisition of private property that was conducted in the Buffer Zone, with the state acting as the 'catalyst for change'. Yet, due to the lack of a convincing feasibility study or comprehensive plan and the set in motion of an expropriation process mired in legal and administrative issues, private sector financing failed to materialise. The governorate's ambitious vision failed to deliver upon its expectations. Besides, in 2014, with the conflict with ISIS and the internal displacement it triggered, and coupled with the fall of the oil prices, the financial and security circumstances of the whole Kurdish region and country have also changed, requiring new innovative approaches to the financing of urban projects to be explored.

6.5 An Emerging Vision for Erbil's Historic Quarters

Following consultations with locals and field assessments during the visioning session *'what would you like Tajeel to be in 20 years' time'*, MOMT staff and colleagues from Erbil Municipality acknowledged that the current building freeze will ultimately lead to the irreversible loss of all historic buildings and urban fabrics. The idea of giving developers and citizens free rein was strongly rejected. Finally, there was a general consensus that life and culture—integral component of Erbil's economic and social identity—should be brought back to the area by realising the full potential of its historic buildings and intricate urban fabric. A sense of appreciation emerged within the group for a closely-knit urban fabric that offered a distinctive level of densification and mixed activities and contributed to a socio-economic performance that modern districts in Erbil appear to have failed to achieve. The result of the visioning sessions was encapsulated as follows:

> **A Vision for Tajeel**[11]
> Local authorities and stakeholders endeavour to achieve:
> A vibrant neighbourhood that realises the full potential of its proximity with both the Erbil Citadel and the busy Qaysari bazaar area, its cultural heritage and its mixed-use vernacular fabric to achieve the wider aims of socio-economic regeneration, and offers residents, the business community and visitors a good quality of life, along with improved key public services, walkability and safety at all hours.

6.5.1 Realising the Full Potential of Tajeel

Ideas regarding the physical design of spaces revolved around two main aspects: quality of the built environment and enhanced open spaces:

(a) Quality of the built environment

- Preserving architectural identity and traditional building typologies.
- Ensuring the preservation of existing typologies, mass and architectural details while avoiding homogenisation.
- Encouraging mixed land uses and housing diversity, catering for different household units, activities and types of visitors.

---

[11] Ideally, the visioning exercise should have also included group of residents and the *mukhtar* (area representative), but because of the sensitivities surrounding the expropriation process and the fear that the exercise might have been misinterpreted as a criticism of the leadership or sparked expectations among citizen, it was decided to limit any further public discussion.

- Set obstacles to storage and warehousing activities, in favour of small-scale commerce, housing and hospitality activities that can ensure street life ('eyes on the street').
- Introducing a few 'iconic' modern buildings in harmony with surrounding heights, materials and courtyard typology.
- Offering incentives for the redesign or makeover of selected eyesore buildings.
- Supporting the creation of courtyard and rooftop cafés that offer a privileged view over the citadel
- Attracting cultural and creative industries (vocational schools, performing arts, laboratories, exhibition spaces, cultural events, theatre, music, etc.).
- Encouraging a culture-led diversification of the economy, through investments on cultural infrastructure and creative enterprises, construction craftsmanship, training in the hospitality industry and new technologies.

(b) Enhanced open spaces

- Recreating walkable spaces that can offer a safe environment for people to meet and gather, for children to play safely and the elderly to sit and chat in tranquillity.
- Marking the entrances to Tajeel by creating landscaped 'wedges'.
- Create an elevated and shaded 'promenade' that connects all the pedestrian streets along Bastay Tajeel(canal) Street and provides much needed public space and protection from the traffic below.
- Discourage traffic in the area by providing multistorey parking facilities outside the city centre.

6.6 An Integrated Management Plan and Implementation Strategy

The desire of governments to try to modernise deteriorated urban centres is understandable. Regrettably, top decision-makers are often led to think that only new and 'modern' housing solutions are worthy of consideration—particularly when areas have entered a vicious circle of decline with under-utilised housing stock. Yet experience has shown that *'almost everywhere, the historic city centre represents a unique historical link with the past, a physical manifestation of the social and cultural traditions which have developed to give the modern city and society its meaning and character'* (Steinberg 1996, pp. 465).

6.6.1 Strategising Interventions and the Leveraging of Land Value

Having 'thrown the baby out with the bath water' by relocating all residents, authorities must now take bold decisions. It appears very urgent to develop a robust planning and area management strategy to galvanise concrete actions from the public and private sectors that can reverse the abandon of the historic quarters to fully reintegrate them in the city and contribute to the growth of the region. It appears crucial for the authorities to take a strong leadership role in the 'fixing' of the situation in the historic quarters. The government needs to take the responsibility to save what social fabric is still left, rebuild trust of citizens by developing innovative solutions to bring back residents, and create an 'enabling environment' for small-scale private investors—with initiatives that should range from reinforcing the existing legal framework to scheduling the urgent upgrading of the infrastructure and services.

Ideas on what would be the most viable strategies for implementation revolved around a suggested mix of intervention approaches that hinge upon the *long-term lease of selected parcels* of land to *cross-subsidies* the restoration of key historic buildings, the *adaptive reuse* of heritage buildings, attracting *small-scale investors* through open but *conditional construction bids*, the *consolidation* of minute and unsellable parcels, the strengthening of *building control mechanisms and tools*, encouraging *mixed-uses* and helping the set-up of *small-medium enterprises* compatible with residential areas to attract a variety of users, customers and visitors (hospitality industry, start-ups enterprises, IT services, arts and crafts, etc.).

The first phase would focus on kick-starting the initiative by *securing political consensus* and project backing among highest authorities, *phasing out the expropriation process*, sharing intent with potential partners and counterparts, setting up communication channels with citizens and identifying *'influential champions'* of the project.

The next step would entail securing a Seed Fund from the Governorate that will allow the Municipality to set up a team of technical experts, tasked to conduct detailed building surveys and GIS mapping of infrastructure conditions drawing upon the expertise of HCECR, conduct feasibility/land market studies, in parallel to building institutional and management capacity at the local level, and setting off community consultations through which to prioritise needs in infrastructure development. The scheme depends on being able to strengthen and bolster the role of Municipality One and its Buffer Zone Committee to oversee the team in charge of the project.

Critical to unlocking the current financial impasse would be a new set of procedures that allow the reinvestment of earmarked revenue from government property in a purposely set up revolving Buffer Zone Fund. The lease of non-historically, less culturally or architecturally significant buildings but with higher real estate potential could be publicly auctioned to raise sufficient capital to kick-start the process. The revolving fund would cross-subsidise the restoration of highly visible heritage buildings and priority infrastructure works, services and utilities through the phased lease/sale of selected parcels of commercially viable land currently owned by the government and part of the Municipal revenue from commercial taxes. Landmark heritage buildings reused for activities of public interest would provide 'anchors' of the rehabilitation project. These components need to be considered in an integrated manner to ensure the implementation of a sustainable project in the long-term.

6.7 Conclusion

A project approach that is more sensitive towards Erbil's urban heritage, receptive to the needs of citizens and that also attempts to deal with the legal framework and existing financial constraints is definitively possible. The rehabilitation of historic quarters such as Tajeel should not ignore their core identity, determined by decades of socio-economic and cultural overlays and enrichment. It appears timely for the authorities to rethink their approach towards city planning and urban regeneration. A public dialogue on the future of Erbil that includes the voice and needs of its citizens would go a long way in enhancing sociopolitical reconciliation and consolidating governance in cities.

References

Al-Hashimi FW, Bandyopadhyay S (2015) The persistent element in the old urban fabric, Erbil Bazar area. J Strategic Innov Sustain 10(2):49

Al-Jameel AH, Al-Yaqoobi DT, Sulaiman WA (2015) Spatial configuration of Erbil Citadel: Its potentials for adaptive re-use. In: Proceedings of the 10th international space syntax symposium, No. 039. http://www.sss10.bartlett.ucl.ac.uk/wp-content/uploads/2015/07/SSS10_Proceedings_039.pdf. Accessed Oct 2016

Dolamari M (2016) History of Kurdish Jews in Erbil, Kurdistan. K24 (Video Documentary)

Foster P (2010) Historical Beijing quarter to be destroyed. In: The Telegraph www.telegraph.co.uk/news/worldnews/asia/china/7532375/Historical-Beijing-quarter-to-be-destroyed.html

HCECR (2014) Nomination of Erbil Citadel Kurdistan for Inscription on the UNESCO World Heritage List, vol I

Ibrahim R, Sabah M, Abdelmonem MG (2014) Authenticity, identity and sustainability in post-war Iraq: reshaping the urban form of Erbil City. J Islamic Architect 3(2). https://www.academia.edu/9844513/AUTHENTICITY_IDENTITY_AND_SUSTAINABILITY_IN_POST-WAR_IRAQ_Reshaping_the_Urban_Form_of_Erbil_City. Accessed Nov 2016

ICOMOS (2010) Erbil Citadel (Republic of Iraq)—No. 1437. Advisory body evaluation

ICOMOS (2014) Evaluations of nominations of cultural and mixed properties to the world heritage list—report to the world heritage committee, 38th ordinary session, Doha, June 2014. Accessed Oct 2016

Khasro AO, Sumarni Il, Daniel JF (2016) The heritage values of Arab district, Erbil City, Iraq. Int J Eng Technol Manag Appl Sci 4(2). http://www.ijetmas.com/admin/resources/project/paper/f201603051457166007.pdf. Accessed 7 Nov 2016

Khoshnaw DS (2016) Gentrification and displacement process: a case study of Erbil. Authorhouse, Bloomington USA

Michelmore D (2013) Rediscovering the city. The Middle East in London, vol 9(12)

Mohammadi R (2014) Moving a market: impacts of heritage nomination on a local community. A case study of Delal Khaneh in Iraqi Kurdistan. MA Thesis, University of Waterloo. https://uwspace.uwaterloo.ca/bitstream/handle/10012/8363/Mohammadi_Rojan.pdf?sequence=1&isAllowed=y. Accessed 12 Oct 2016

Monk D, Herscher A (2015) The New Universalism: refuges and refugees between global history and voucher humanitarianism. Grey Room 61. MIT

Neurink J (2014) At Erbil Citadel, UNESCO Wants to See More of Ancient Past. Rudaw. http://rudaw.net/english/kurdistan/060620141. Accessed 15 Nov 2016

Oz O, Eder M (2012) Rendering Istanbul's periodic bazaars invisible: reflections on urban transformation and contested space. Int J Urban Regional Res 36:297–314

Steinberg F (1996) Conservation and rehabilitation of urban heritage in developing countries. Habitat Int 20(3)

UNESCO (2014) "Erbil Citadel". In: World heritage list. https://whc.unesco.org/en/list/1437

Watson I (2007) Kurds displaced in effort to preserve ancient city. NPR, Feb 4, 2007. http://www.npr.org/templates/story/story.php?storyId=7133379#714066. Accessed 3 Oct 2016

Chapter 7
Silk Production as a Silk Roads Imported Industrial Heritage to Europe: The Serbian Example

Milica Kocovic De Santo, Vesna Aleksic and Ljiljana Markovic

Abstract This chapter investigates industrial heritage related to silk production in Serbia. Silk and silk processing craft were imported from Asia through Silk Roads (across the Balkans and Italy towards Europe). Silk production on industrial level started in today's Serbia in the middle of the 18th century. The whole process of silk production initiated the growth of mulberry trees, that changed cultural identity in the sense of the 'everyday way of life' and broader application and use of wood plant even today. Mostly based on our archival data, we mapped cities that had silk production in Serbia, highlighting interrelated fields such as culture, economy and identity, initiated by silk production. We emphasised the importance of imported silk-related phenomena and common cultural heritage, but also the authentic cultural identity that came out from the silk production in Serbia. It was important to examine all relevant institutions and sectors involved (historical perspective) around the silk production in Serbia, with consideration of possible alternative ways of integrative management options in order to govern and make industrial silk heritage more visible and valuable. Our focus and proposal in this sense go to the direction of participatory governance options, which will enable the multi-departmental approach. Our main research questions are:

I. How important are the phenomena that came from Silk Roads seen as common but also authentic heritage?
II. How to create a new system solution (for use, managing, conservation and risk mitigation of silk (roads) heritage (creation of new alternative (touristic) products and better local integration, regional, international)?

M. K. De Santo (✉) · V. Aleksic
Institute of Economic Sciences, Belgrade, Serbia
e-mail: mickocovic@gmail.com

V. Aleksic
e-mail: vesna.aleksic@ien.bg.ac.rs

L. Markovic
Faculty of Philology, Belgrade, Serbia
e-mail: ljiljana.markovic@gmail.com

© Springer Nature Switzerland AG 2020
F. F. Arefian and S. H. I. Moeini (eds.), *Urban Heritage Along the Silk Roads*,
The Urban Book Series, https://doi.org/10.1007/978-3-030-22762-3_7

III. What type of new system solution/management model is optimal for achieving optimal integration in relation to heritage-local people-values?
IV. Why do we propose participatory governance with the aim to increase synergy and involvement of stakeholders?

Keywords Industrial silk heritage · Participatory governance · Mulberry tree cultural identity · Alternative solutions for commons

7.1 Introduction

Silk production method(s), its origin and the way in which silk arrived in all parts of the world are still important and interesting from many perspectives. The sericulture industry is determined by the man–insect–tree–factory relation. Bombyx Mori is the most commonly used kind of the silkworm, which passes through several stages: from seed to the cocoon onto the transformation into a butterfly. The insect is considered a domestic animal because it could not survive without human involvement. When breeding silkworms, it is necessary to cultivate mulberry trees, because the silkworms feed on mulberry leaves. This is the simple relation about the breeding of Bombyx Mori, man and a mulberry tree.

Sericulture arrived from the Far East to Western lands. In this sense, our research contains an interdisciplinary approach to correspond with links between culture, history and economy. In fact, the silk is very important because it summarise and syntetigese the meanings and values of cultural, historical, economical and industrial heritage (Kocovic De Santo, Markovic, De Santo 2018). We give brief historical facts related to the intercontinental silk production movement, whilst in the last part of this chapter, we focus on the Serbian example of silk production and its challenges. Based on our findings, possible future recommendations related to silk governance will be also discussed.

7.2 A Short History of Silk's Intercontinental Movement and Production

It is widely believed that the oldest examples of silk date from the period 2850 to 2650 BC (Northern China (Federico 1997), the valley of the Yangtze River (Hansen 2012)). These findings mostly came from archaeological evidence. Hansen (2012) notes that the Chinese were indeed the first people in the world to make silk as early as 4000 BC. Moreover, the author notes that 'according to Hangzhou Silk Museum, the earliest excavated fragment of silk dates to 3650 BC and is from Henan Province in central China' (Hansen 2012). Previous examples are often taken controversially and sceptically by non-Chinese experts, so we can conclude that there is a globally accepted consensus at least about a time between 2850 and 2650 BC for the emergence of silk.

Even if we take as legitimate any of the above mentioned dates, there is a lack of information about silk production. Furthermore, it is difficult to speak about silk production at an industrial level for that time. When Federico (1997) speaks about the beginning of silkware trade from China all along the Silk Roads to Far West (Greece and Rome), he points out that in early days 'it was so expensive to make long-distance trade profitable'. This would be clearer if we imagine all the logistics and transportation options of the time that mostly relied on animals through passive and not very amicable landscapes, often without clearly marked routes, from the deserts to the highest mountain peaks. If we want to illustrate the adventures undertaken by traders and travellers, it is good to know the facts about Charles Blackmore 1993 expedition through the Taklamakan: 'His men and camels were able to cover a distance of 59 days for 1,400 km through the desert of Loulan to Merkel, southwest of Kashgar, which means that, on average, exceeded 21 km per day. Marching over sand dunes in the desert was so hard that they could not proceed more than 16 km daily, but they are managed to move up to 24 km when they walked on the flat, gravelly soil' (Murray 2000; Hansen 2012).

Let us return to the legend, which says that the princess smuggled silkworm. How would the world look like today, if it did not happen? When and whether the mulberry tree would reach Europe? According to previous research that started from the late nineteenth and early twentieth centuries, we can identify a few routes of smuggling silkworms. Sericulture actually reached many areas through different channels. Around 200 BC with Chinese migrations, it came to Korea. Federico (1997) mentioned that in some Indian literary sources from 1300 to 1400 BC silk production was already mentioned and familiar in India. It might have been a different kind of silkworm, the wild type that produced tussah silk. In his book about the economic history of the silk industry 1830–1930, and with reference to previous mainly anthropological and archaeological researches about silk expansion Federico (2009) writes:

> The next wave of expansion of silk production began around AD 300–400. Within China, sericulture moved southwards to the Central region (around Shanghai), which was to become the country's main producing area after the eleventh century. Silk production began around the same time in Japan too, where it was almost certainly imported from China, the nationalistic legends notwithstanding. In AD 552-556 sericulture reached the shores of the Mediterranean. The silkworm eggs were smuggled out of Persia by two monks on a mission on behalf of the Byzantine Emperor Justinian. Further diffusion of sericulture westward was halted for some centuries, probably more by the poor economic conditions than by the attempts of the Byzantine emperors to keep a monopoly on it. Eventually, the Muslims brought sericulture to Spain (in the ninth century) and Sicily (in the eleventh). From there, it began a slow march northwards along the Italian peninsula, spreading all over the Northern regions in the fifteenth and sixteenth centuries. It arrived in France in the late fourteenth to early fifteenth century, but large-scale silk production began only some two centuries later,

as was the case in Austria. At the same time, sericulture resumed its movement southwards in China, also localising around Canton. Nearly 5,000 years after its first discovery, the production of silk had spread to most of Asia and to the whole of Southern Europe. Its worldwide location was not to change for the next three centuries, despite an enormous growth of production.

Along the silk roads, numerous goods were exchanged, in different ways, and among different cultures, languages and people, and that is why these roads could be seen as one of the first transnational and global corridors that provided migrations of vibrant values through culture and economy. This is important because of the fact that a long history of different cultural exchanges brought many (reshaped) universalities. There is a wealth of historical evidence about long traditions of cultural and goods exchange throughout millennia—yet no factual knowledge about the meaning and significance of the relatively new phenomena such as global, transnational, cultural and economic. Besides the silk, parallel exchanges that took place on these roads (languages, religions, literature, writings, scripts, spices, legends, materials, grains, animals, people, etc.) enabled the great cultural mix and exchange. In this sense, Silk Roads are a planetary phenomenon due to the strong cultural domino effect.

7.3 Silk Industry in Europe: On Globalisation and the Serbian Example

As mentioned, silk production on an industrial level in Europe started in line with the modern industry, but also in some examples it had its base in proto-industrial phase. In some silk production cases labour was flexible in working in agriculture and industrial departments, which by implication means that silk and silk production allowed the creation of social, financial and cultural linkages links between urban and rural parts (Kocovic De Santo, Markovic, De Santo 2018). Global proto-industries (in archaic globalisation terms) contributed to the development of the world silk market. Previously, silk was mostly produced and traded on a domestic level. The paradigm begins to change with the emergence of the growing demand for silk. According to Federico (2009), in the middle of the fifteenth century, the French industry thrived so much that the national production of raw material remained constantly insufficient to needs. Many of the faster developing industrial northern countries could not produce larger amounts of silk mostly because of climate and economic challenges. What came as a result was the import of Mediterranean and later Indian, Chinese and in the middle of nineteenth-century Japanese silk. Federico also mentions few imports from the Balkan region (Bulgaria, Yugoslavia, etc.). The problem with the Balkan region was that the purchase of raw silk was conducted by private cartels and exported without labels to European silk markets (Lyon, Milan, UK, etc.) (Federico, 2009). The silk from the

Balkans was mis-sold wholesale as Italian or Hungarian. As Federico notes in a 100-year period from the 1820s to the 1920s there was a 20-fold growth in quantity. From the 1870s to 1929, the value of trade increased nine times, and the world trade/output ratio increased from 0.5 to 0.75. It is interesting to mention the evolution of silk status as a prestigious luxury good into an increasingly accessible good for everyone. The use value of the silk had many variations, from fetishistic artisan or native etc. At one stage it even developed into a substitute for money as a means of payment and the credit guarantee (Kocovic De Santo, Markovic, De Santo 2018). The growth was abruptly interrupted from the 1930s to the 1950s. This was due to negative effects of the Great Crisis and World War II, but also the invention of several artificial substitutes such as oil-based polyester causing a decreased world consumption of silk of more than 60% of the levels reached in the 1920s. The effects were uneven, with sericulture disappearing altogether from Europe and the Middle East and the Japanese production falling by two-thirds. China remained the only exporter of silk.

Previous trends in world silk production effectively represented the mainstay of globalism and development, which included progress in all domains of human activity and life 'without borders'. In the decades to come, we could realise that globalisation in many cases favoured exogenous development (induced or stimulated by external factors) compared to the endogenous (development based on internal resources specific territory). This difference sometimes led to the disappearance and/or major changes in the authentic cultures and economies. Seen in this sense, we can say that globalisation was one of the 'soft' reasons for a future of decline and disappearance of silk production in Europe, Balkan, Yugoslavia and Serbia. There is, of course, a variety of interpretations as to whether or not the silk production should be seen as highlighting commonalities between nations.

7.3.1 Silk Production, The Serbian Example

As mentioned above, silk-like fabrics arrived in Europe from Asia, through the East Roman Empire (Byzantine) and the Mediterranean. This happened over a period of hundreds of years. Silk brought its associated cultures and economies. Culture and economy, with the arrival of silk, were formed and reshaped within any terrain in contact with the indigenous. The silk production as an (economic and cultural) activity combines traditional and imported knowledge (Kocovic De Santo, Markovic, De Santo 2018). Therefore, Chinese or universal, sericulture has been very authentically developed in each place. It is likely that the mulberry tree, would not arrive in Europe if there was no sericulture. This can be claimed with even greater certainty when talking about the white mulberry tree, necessary for the cultivation of silkworms. However, many argue that the black mulberry was indigenous. The oldest black mulberry tree is still alive in Pec Patriarchate (one of the most valuable Serbian heritage and monastery from UNESCO World Heritage list in Kosovo) and it is 700 years old.

7.3.1.1 The First Phase of Sericulture

The first factories were initiated in the mid-eighteenth century, including Panchevo, a city in Banat, northern Serbia. We can consider this as the first phase of sericulture in Serbia. At that time Banat was within Austrian military borders. Banat had a strategic and political role. According to Milleker (1926):

> Banat border was another link of the chain of military Austrian border, which covered the territory of the Adriatic Sea to the borders of Bukovina in its highest prevalence in 1851. In the time of Habsburg monarchy it constituted a defensive rampart for the old enemy and the plague, which has also been deadly. Culturally (religion and language), the population was highly diverse, but the Serbian population has been present here since the 15th century retreating from the Turks.

The production of silk was introduced in 1733 by Count Mercy, and rebuilt in the Banat border in 1769, at which time the production amounted to 528 lb of cocoons. The Unwinding Silk Bureau was founded in 1776, in the city Bela Crkva (White Church) and Panchevo. According to Đukic (2017), 'the industry in this period was based on the processing of agricultural crops on the fertile land of Banat. Thanks to the cultivation of mulberry tree sericulture was introduced in 1733'. As Đukic (Đukic, 2017) notes 'mulberry trees have built "the soul of many cities" in the past, allowing the production of silk, homemade brandy (rakija) and jams from the fruit of mulberry' (Ibid.). This is a very important aspect of the diverse use value of mulberry trees. In this sense, the mulberry tree purely represents imported cultural phenomena that still exist in Serbia as a dissonant heritage with layers of many added cultural values. Even if today's young generations do not connect mulberry tree with silk anymore, (because the trees are almost completely exterminated compared to the hundreds of thousands of planted trees in the eighteenth and nineteenth century) these cultural aspects are highly important for the future.

7.3.1.2 The Second Phase of Sericulture

We can say that the second phase of sericulture that is important for the Serbian case begins after WWI (1914–1918), and later the establishment of the Kingdom of Serbs, Croats and Slovenes. For this phase, we mostly used information from the Historical Archives of Yugoslavia, as we found it most relevant. What makes this phase significant is the fact that it coincides with parallel sericulture development periods elsewhere in Europe (and also Japan and China) in the sense of global silk markets, but also because sericulture and Silk Roads were peaking in this period.

During the interwar period (World Wars I and II) sericulture industry in the Kingdom of Yugoslavia (KY) was divided into state, concession (private) and free-range sectors. The financial support for the development of the sericulture was mainly received from the state subventions, but also from private banks whose capital was dominantly foreign (Aleksic, Malovic 2017).

State-owned sericulture stretched on the territory of Vojvodina, Bosnia and Herzegovina, Dalmatia and Montenegro. All the factories were state-owned. Also, as mentioned, sericulture in the Kingdom of Yugoslavia was promoted as a social and not just profitable category. The intention was to give the opportunity to the most vulnerable people (poorest agrarian communities) to earn more. By breeding silkworms, they could get additional income before any annual agricultural income is received (the cocoon selling season was June–July, before harvest time). Put in a contemporary language we would say that Government was being socially sensitive through strongly supporting public entrepreneurship. Previous this is especially correct until the outbreak of the Great Economic Crisis, which arrived in the Kingdom of Yugoslavia in 1931 (Gnjatovic, Aleksic 2011). Therefore, the entire organisation of the state sericulture was directed towards planting and cultivation of mulberry trees. Mulberry trees were mostly located on common and public areas, and the planting process was supported by the state budget. Mulberry trees were free for farmers to use for leaves collection in the process of silkworms breeding. The state allocated free mulberry saplings and silkworm eggs for everyone, and buy (for further processing as the end-product fabric) cocoons at prices that were in line with the world market. Throughout the KY there were 24 stations for the purchase of cocoons, 10 nurseries with around 300 thousand pieces of mulberry under five years of age, and even 700 thousand mulberry trees planted along streets and public roads. At the same time, more than 25 thousand families were engaged in sericulture. The head of state sericulture was the State Department of sericulture (DS) in Novi Sad with a mission to promote the industry in the entire KJ territory. With over 15 of DS jurisdictions, nine were supplied with famous Pelegrino stoves for 'suffocation while drying cocoon' needed for silk processing. Each of these jurisdictions represented a kind of regional silk centre that included 20–40 municipalities.

There were only two concession companies for the cultivation of silkworms. One of them was the Croatian agricultural bank that covered the whole area of Croatia, Slavonia and Međumurje (also parts of today's Croatia). The second was the Silk association located in Lapovo (Serbia) that covered the pre-war area of Serbia. Free-range breeding of silkworms existed only in South of Serbia within the KY (today's Macedonia).

The progress of sericulture in the entire country was hampered mostly by the fact that two Ministries were formally in charge of this industry. The Ministry for

Agriculture and Water (MAW) was responsible (through the body Central Silk Management) for the part related to the production of mulberry trees and purchase of cocoons from producers. The Ministry of Trade and Industry (MTI), was responsible (through the body Public Silk Factories) for further industrial processing and sale of silk. The challenges arose because the cocoon price decided in one ministry (MAW) represented the price for the final product in another ministry (MTI). Frequently occurring errors or delays in the first stage led to the selling of silk without having full balanced information. Also, the inability to complete the process of monitoring the production cycle led to lower output levels than planned. The costs of production could not be reduced in time, which put the state and producers always in the position of material loss. The Council of Ministers of all government ministries kept discussing possible directions until 1924 when sericulture got unified management under the MTI. The second challenge, perceived by MTI lied behind the fact that in certain areas sericulture was hampered by concessions operations. According to the Minister's view, concessions were slowing this industry due to holding a monopoly in the purchase in some parts of the country and dealing with market speculation on internal, but also external trade. The concessions were engaged in foreign trade. As private citizens, they were not interested in the product and greater social benefits from the Yugoslav silk but led by personal profit, selling the state-made silk unmarked claiming that it was produced in Italy or Hungary on the silk markets abroad. After the unification of sericulture industries under one MTI, the Kingdom of Yugoslavia started to export silk very successfully on international markets mostly to Lyon, Milan and Zurich.

In 1928, silk production in the Kingdom of Yugoslavia was in the third place in Europe with high-quality types of cocoons, due to the suitable weather conditions that contributed well to sericulture development (Archive Yugoslavia, Memorial book 10 years of KSHS 1918–1928). Practically, there was not a single municipality in the Vojvodina (North of Serbia), which were not breeding silkworm. This resulted in the recognition of sericulture as one of the most important state industries in the KY. Only in North of Serbia (Vojvodina) where the three largest factories were located was it possible to produce annually million to million and a half tons of raw cocoons, which represented around 400–500 tons of dry cocoons.

The three pre-World War II main silk factories are shown in Table 7.1. We are not sure what damage this industry had suffered, due to a lack of information (since the National Library in Belgrade was bombed during WW II). All three factories have annually been able to produce between 38 and 40 tons of raw silk. All the products and related items produced in state factories were exported abroad, mostly to Milan, Zurich and Lyon. Due to its excellent quality, the product in interwar Europe was highly valued, even with final prices above common gross stock prices.

Table 7.1 Main silk factories and annual production before IIWW

Factory	Number of workers	Production of cocoons	Silk production
Novi Sad city	350–400	60 tons of raw cocoons/ annually	15 tons of raw silk/ annually
Pancevo city	300	50 tons of raw cocoons/ annually	12 tons of raw silk/ annually
Nova Kanjiza city	220	45 tons of raw cocoons/ annually	11 tons of raw silk/ annually

Source Authors table based on the data from Archive of Yugoslavia

7.3.1.3 The Third Phase of Sericulture

This phase begins in 1946 (after the end of World War II), when the country has suffered huge human and material damage and the Federative Republic of Yugoslavia (FNRY) was established. Ideologically different, the new-born socialist country began to develop a planned economy, emphasising the progressive development of heavy industries (metallurgy, mining, etc.). The concentration on heavy industrialisation sidelined even light industries (such as textile). Agriculture ran out of steam in the next few decades, with its value allocated to the development of heavy industries.

A key factor to take into account was the emergence of so-called 'artificial silk' whose fibres had a certain similarity with natural silk but obtained by chemical procedures that included petroleum. There was a new global paradigm of mass petroleum use, started from the USA. This situation slowly pushed the force of natural silk not so much in quality than in price. Still, natural silk was found necessary for certain products (parachutes, mill sieves, insulating materials for electrical industries, etc.).

Sericulture found its place in the first 5-year economic plan of socialist Yugoslavia. The plan found scope for the expansion of silk. From the 460 tons of silk cocoons, produced in 1946, it was considered very easy to increase production to a million and a half or even two million tons in the period between 1947 and 1951. Within the first five-year plan the development of the two million tons per year production was designed. The same five-year plan allowed for the construction of two facilities for final products of silk, with a capacity of about 266 tons per year. One large Installation containing a capacity of about 180 tons per year was provided in Titov Veles (Makedonija) while the silk factory in Novi Sad was provided to increase the capacity to 52.5–55.5 tons.

We could not trace the information about silk and economic plans from this phase, but we can see new trends towards chemical industries worldwide, replacing almost all natural and clean industries until the 60s. After 1960, the silk ceased to exist on an industrial level in Yugoslavia.

7.4 Findings and Recommendations

This chapter included our findings of sericulture. First of all, sericulture is a combination of endogenous and exogenous development approaches: exogenous because it was imported and endogenous because it combines traditional and imported knowledge in the processes and, management practices. Moreover, the organisation of sericulture work implies the linking of agriculture and industry, or from labour perspective farmers and workers. Still, it is a combination of endogenous and exogenous values that link the process of silk production. The innovative creative economies and nature conservation, brought awareness that endogenous and exogenous cultural and economic developments are very important (Đukic 2017). Future planning and governance in order to restart sericulture should include integrative approach in order to manage endogenous and exogenous aspects related to silk production.

The second important point is that sericulture represents a dissonant heritage, whose interpretations are tightly influenced by cultural memories and identities of interpreters. According to Sesic and Rogac (2014),

> … the experiences of the present are largely based on specific knowledge of the past. Thus, the ways of experiencing the present are influenced by different perceptions of the past with which it can be connected (…) all the heritage is a contemporary interpretation shaped by narratives of history(…). Dissonant heritage often opens up ambivalent topics and largely unwanted past through the narratives which is the right key for future management solutions for silk heritage re-activation. This is also an important aspect and argument for sericulture revitalisation, as it represents a confrontation space. But also, silk and silk roads are attributed as common goods. As such, in terms of laws, and the degree of use and exploitation these types of property are mostly recognised to be managed by the community: bottom-up approaches, with strong inclusiveness (Kocovic De Santo, Markovic, De Santo 2018).

The third important finding is that sericulture in the Kingdom of Yugoslavia was important, in contemporary terms, as a social entrepreneurship since it was focused on socially responsible jobs, which helped vulnerable groups.

The fourth finding of sericulture reflects the fact that through time it represents the example of clean industry, which makes it desirable even today. Revitalisation and re-establishing of sericulture could have a positive contribution to sustainability which is today the most topical topic and mission to achieve.

Based on these findings, we agreed that the combination of different departments with integrative approach could be the right governance option for silk heritage. In other words, new solution systems for silk could point towards the direction of participatory governance. Participatory governance (PG) occurs as a response to the problems that characterise transitional societies in developing countries. The way in which PG provides its positive impacts on vulnerable societies' shocks is through higher transparency, active multisectoral cooperation, fair distribution and greater inclusion of local people (Kocovic and Djukic 2015). Kocovic (2017) also proposed a new solution system for managing heritage, in order to make it more valuable. This solution is suitable for sericulture because it actually connects alternative forms of tourism—that are seen as contributors to the sustainable development of heritage. New solution system for managing silk heritage in order to achieve sustainable development and new creation of values should include three recommendations. The first is related to management option of the participatory governance of silk heritage with linking strategy and partnerships of relevant stakeholders.

The second is related to the creation of new products that rely on silk heritage. This could be achieved through alternative forms of tourism that use an integrative approach for dealing exogenous/endogenous know-how around dissonant industrial silk heritage. The third recommendation refers to the risk management of heritage and visitors, but also visitor management, in order to maintain the safety of silk heritage as well as people (Kocovic et al. 2016, 2017). This means that alternative forms of tourism (such as cultural), are seen as main contributors of silk heritage, but also silk production revitalisation. Since tourism relies on narratives, (similar to dissonant heritage) in the creation of new products, these narratives should be better connected with narratives around silk.

Finally, if sericulture as a clean industry had been seen at the beginning of the twentieth century as a socially responsible field of work which, that created a lot of jobs for different social groups, we can conclude that the re-establishment of this industry would mean a positive contribution to sustainable development in general, but also sustainability and preservation of the silk industry as an imported heritage.

References

Alesić V, Malović M (2017) Past development of Serbian enterpeneurship: the case of privately-owned banking corporations. In: Radović Marković M, Nikitović Z, ZANADU LC (eds.). Insights and potential sources of new entrepreneurial growth: proceedings of the international roundtable on entrepreneurship 4 december 2016. Belgrade. Bologna Italy Filodiritto Publisher Inforomatica pp 42–54

Archive of Yugoslavia documents: MTI, Number of funds: 65, AJ 1116, AJ 958 Number of folders 65; 424; 328: Silk price on world market 1922, Law for State silk factories; Memory book: 10 years of The Kingdom of Serbs, Croats and Slovenians 1918–1928; The report of the board of Serbian Silk Association; Labour rules for State Silk factories 1920; General financial plan for textile industry 1947 Federative Republic of Yugoslavia. MTI, Number of funds 65, Organization unit: Industry; XI group Textile Industry, AJ 959, Number of folders 329, years 1925–1940 65-329-959: 1926–1945 Silk production documents, (Letter to the Minister 1924 from Central Silk Management; Letter to the MIT from the evaluator and member of steering committee from State Silk Factories 1924; Letter of Minister MIT to the Ministry Board 1925; Regulation on the commercialization of the national directorate of sericulture and silk factory district 1933

Charles B (2000) Crossing the desert of death: through the fearsome Taklamakan. John Murray, London, 59, 61, 64, 104, caption to f g. 14

Đukic V (2017) WHO we are—how we are: a study of memory and identity policy in Serbia, Institute for theatre, film, radio and television, Faculty of Dramatic Arts, Belgrade

Federico G (1997, 2009) An economic history of the silk industry, 1830–1930. Cambridge Studies in Modern Economic History, Cambridge University press, digitally printed version

Gnjatović D, Aleksić V (2011) Rescuing Agricultural and Banking Sector from Collapse: Agricultural Debt Consolidation in Yugoslavia 1932-1936 Industry J Econ Inst – Belgrade 2, pp 283-297

Hansen V (2012) The silk road, a new history. Oxford University press

Kocovic M, Djukic V (2015) Partnership as a strategy to achieve optimal participatory governance and risk mitigation (of cultural and natural heritage). ENCATC J

Kocovic, Milica, Djukic, Vesna, Danijela V (2016) Making heritage more valuable and sustainable through intersectoral networking. ENCATC J Cultural Manag Policy, an online magazine

Kocovic M (2017) Thesis: the contribution of eco-cultural tourism to a sustainable development of protected areas with associated cultural and natural heritage. Faculty of Dramatic Arts, Belgrade

Kocovic, Jovicic, Babic (2018) Challenges and alternatives related to the financial analysis and management options for achieving sustainable development of protected areas and commons. Mag Theme 42(3):939–959

Kocovic De Santo, Markovic, De Santo (2018) Silk (as an Imported heritage) and it's market evolution in the South of Italy, Calabria example. J Int civilization stud 3(1):153–173

Milleker M (1926) History of the Banat of military border 1764–1873 (pp 10). reprinted Pančevo History Archive 2004

Sesic M, Rogac ML (2014) Balkan 'Dissonant Heritage Narratives (and Their Attractiveness) for Tourism'. Am J Tour Manag. https://doi.org/10.5923/s.tourism.201402.02

Part III
Post-War Reconstruction and Urban Heritage

Part 10

Chapter 8
Craftsmanship for Reconstruction: Artisans Shaping Syrian Cities

M. Wesam Al Asali

Abstract Craftsmanship had a crucial role in shaping Syrian society and its relation to the built environment. Organisations and guilds of artisans were official regulators and mediators between artisans, state architects, courts and private house dwellers. It was through these guilds structure that the training of crafts took place to provide technical and social know-how to contextualise those structures within the city. Also, it was through the guild that the technology transfer came about by immigrants and moving artisans. The paper argues that the loss of this medium at the beginning of the twentieth century resulted in a dichotomy between architecture and building practice, specifically by excluding the popular builder from any formal representation in the state. This dichotomy led to fundamental alterations in modern construction, the nature of the planned and unplanned, the local and global and the private and public. Illustrating the development of the guild structure could reveal how the public and private construction developed in the region. The paper rereads two pivotal texts written about craftsmanship in Damascus in the 1880s. It compares the social structure in construction guilds with those of guilds in other disciplines, and explains the distinctiveness of the builders' guild and its relation to the state. Finally, the paper examines the remnants of guilds structures in the modern Syrian society, specifically in the popular non-institutional training systems. In Syrian cities today, architects are not the primary key players and, in some cases, are completely absent. Almost half of modern Damascus and Aleppo is built by builders with a minimum, if any, intervention from architects or planners. Unlike the traditional artisans and their guilds, those modern builders have no official representation. How could any rebuilding plan be more inclusive and address those builders? There could be a possible answer in the relationship between craftsmanship and the city.

Keywords Craftsmanship · Reconstruction · Guild · Building trades · Damascus

---

M. W. Al Asali (✉)
University of Cambridge, Cambridge, UK
e-mail: mwa24@cam.ac.uk

8.1 Introduction: The Guild

Craft trades were part of the socio-economic character of major Syrian cities, guilds had a crucial role as channels of representation for artisans in their societies and regulated their relationships to their states. In Damascus 1850–1900, notes on builders' guild show the essential role of master builders in regulating micro-planning, working directly with residents and authorities, and participating in private and public construction. This suggests understanding craft as a model that is practice-and not just object-oriented, a model with which might have lessons to learn for the reconstruction of Syrian cities.

Early writings of orientalists loosely use the title 'Islamic guild' to refer to any form of artisans or trade organisation in the middle east.[1] In the current literature, however, the term is better defined and allocated to specific periods, such as the early Seljuk or Ottoman, during which the presence of artisan formations was financially and socially effective.[2] The term guild is a translation of many Arabic names of artisan's groups such as *esnaf, sinf, hirfa* and *tā'ifa*. The latter was the most used in Syria, literally meaning grouping, describing not only artisans but also religious, racial and social groups. The socio-economic aspects of the guild in the history of the Middle East are studied in detail. Current scholarship dives in micro cases of primary resources to look at the economic, religious and social diversity in artisans' societies. Scrutinising the same sources from an architectural viewpoint to find construction-related data could enhance our understanding of how artist-builders related to their cities and forms of authorities.

There are fewer sources to firmly describe if and how artisans grouped before the fifteenth century.[3] Guilds as a medium and regulator of work are believed to have no substantial existence in pre-Ottoman eras where the production was simpler, the population was smaller, and governing was more centralised. The supervisor of trade *al Muhtasib* controlled the crafts regulations and was responsible for the quality of goods and work.[4] In construction trades, *al Muhtasib* supervised only materials and dimensions of walls, and design or urban related matters was out of his concerns. However, during the ruling of the Seljuk dynasty in Anatolia, some forms of social or labour organisations did exist. *Akhi*, which comes from Arabic 'my brother-brotherhood', was an organisation where merchants met, artisans trained and the youth volunteered. The combination of the three makes *Akhi* closer to a society than to a mere guild.[5] The influence of such corporation is evident in Ottoman guilds and Sufi groups. Both shared similar structures and ceremonial

---

[1] For early writing about guilds see Massignon, 'La Structure Du Travail À Damas En 1927'; Lewis, 'The Islamic Guilds'.
[2] Baer, 'Guilds in Middle Eastern History'.
[3] Baer.
[4] Rafeq, 'Craft Organization, Work Ethics, and the Strains of Change in Ottoman Syria'.
[5] Arnakis, 'Futuwwa Traditions in the Ottoman Empire Akhis, Bektashi Dervishes, and Craftsmen'.

habits that can be traced to the ones in *Akhi* organisations. Also, being mainly formed by immigrant artisans who came from Central Asia to Anatolia, *Akhi* had a close affiliation with moving artisans from different cultures of construction.[6] Moving artisans were also part of the Ottoman guild, specifically in construction trades. *Akhi* corporations paved the way to a knowledge exchange in building trades that continued during the successive dynasties in Syria,[7] such as the Ayyubid and Mamluk.[8]

Unlike crafts organisations before the fifteenth-century formations, Ottoman guilds and crafts, including the ones in Syria, are studied and documented in many sources. Writings of historians or travellers, court documents and documents from High Advisory Council and Ottoman Imperial Orders give a clear understanding of the substantial role of Ottoman guilds in the Damascene society. The organisations to which artisans belonged functioned as regulators of the production process, quality, tax collection systems and legislations. Moreover, it was through the guild that the concept of skill and mastery was defined and craftsmen complied with the rules for training and employment. Training had various structures in different guilds, but there were always three typical levels: apprentice (*al-ajir* or *al-mubtadi*), journeyman (*al-sani*) and master (*al-mu'allim*) (Fig. 8.1). For a journeyman to become a master, he needed, after the blessing from the guild masters, the financial ability to become independent. The ability was conditioned by having a place to work (*khiliu*), and the equipment (*gedik*), both formed part of the guild's monopoly, but they also contributed to demand and production balance.[9] Guild masters selected the head of the guild (*al-shaykh*) before presenting him to the city judge (*al-qadi*) for a symbolic approval. Finally, all guilds had the head of guilds (*shaykh al-mashayekh*) who had a religious and cultural significance rather than an administrative one.[10]

---

[6]Wolper, *Cities and Saints* 75.
[7]Kuban, *The Miracle of Divrigi*.
[8]Meinecke, 'Mamluk Architecture, Regional Architectural Tradition: Evolutions and Interrelations'.
[9]Baer, 'Guilds in Middle Eastern History'; Hanna, *Artisan Entrepreneurs in Cairo and Early-Modern Capitalism (1600–1800)*.
[10]Rafeq, 'Craft Organization, Work Ethics, and the Strains of Change in Ottoman Syria'; Baer, 'Guilds in Middle Eastern History'.

Fig. 8.1 Damascus-Sword Maker circa 1900, the pictures show the three levels of mastery: apprentice, journeyman, master. *Source* Library of Congress. USA

8.2 The Guild of Builders in Damascus 1850–1900

To understand the work of traditional master builders in Damascus, I examined two writings about craftsmanship and guilds in Damascus, both written during the late Ottoman Empire, shortly after the reforms (*Tanzimat*) 1839–1876, known for aiming to centralise the ottoman governmental apparatus. During this period, crafts were about to be in competition against industrialisation, the reforms were finding their way to implementation, and guilds and courts were giving up their planning roles to city councils (*Baladiyat*).[11]

The first writing is *Qamus al-ṣinaʿat al-Shamiyya* (Dictionary of Damascene Crafts) by Mohammad Said al-Qasimi who started his research in 1892. In 1900, al-Qasimi died and his son Jamal al Din al Qasimi and son in law Khalil Al Azem continued the work until the completion, which has no exact dates. Louis Massignon incorporated the book into scholarship and used it for his research about

---

[11]City councils law was initiated in Damascus in 1876.

crafts in Damascus in 1927.[12] The dictionary is an 'encyclopaedic paper' that sheds light on the Damascene crafts—al-Qasimi wandered around counting workers and enquiring about professions in the city of Damascus. The second is the essay of Elias Qoudsi *'Nubdha tarikhiyya fil hiraf al Dimashqiyya'* known as 'Notice sur les Corporations de Damas', presented in 1883[13] in the sixth International Congress of Orientalists in Leide. It is another fieldwork piece that focuses on the life of crafts and trades traditions, habits and ceremonies. Both writings show that the definition of the guild of builders and its relationship to the state master builder, court and architects is too complicated to be framed as just another group of artisans in the city.

While working on al-Qasimi's text, I noticed that he uses the book of Ibn Khaldun in 'Al Muqaddimah'.[14] Al-Qasimi chooses some of Ibn Khaldun's text on building crafts and edits it to comply with the context of Damascus, i.e. the techniques and materials in Damascus.[15] This extensive quotation makes the al-Qasimi's contribution questionable, but the fact that he adopts some parts of the texts and changed other parts to fit his fieldwork research could make the original version of Ibn Khaldun also valid in Damascus 1880s. Under the umbrella of the guild of builders al-Qasimi places different building trades techniques like brickwork, stonework and rammed earth construction, which he explains in later sections as independent crafts (Table 8.1).

So what is a builder in Al Qasimi dictionary? In addition to the required technical knowledge, a builder is someone who can use this knowledge to solve contested matters in the city. '… these matters are clear only to those who know construction in all its details. They can judge these details by looking at the joints and ties and the wooden parts….'.[16] The builder is a reference to building regulations. 'The authorities often have recourse to the opinions of these men, about construction matters which they understand better' in a context with technical and social complexity, where people 'compete with each other for space and air above and below and for the use of the outside of a building'. A builder is, therefore, a problem solver who mediates between neighbours about rights of access and privacy. Indeed, some court records show that builders experience were essential in solving such conflicts between neighbours.[17]

Likewise, the text of Elias Qoudsi has a small passage that is worth further attention. 'The guild of builders and stone masons, who are all Christians, have no

---

[12]al-Qasimi, al-Qasimi, and al-Azm, *Dictionary of Damascene Crafts (Qamus Al-Sina'at Al-Shamiyya)* 26.

[13]Qoudsi, *Notice sure les Corporations de Damas (Nubdha tarikhiyya fil hiraf al Dimashqiyya)*.

[14]Khaldūn, *The Muqaddimah* 320.

[15]For example, Al Qasimi does not quote Ibn Khaldun text on reed and mud construction for barns and houses, which is seen more in Iraq and Egypt than in Central Syria.

[16]al-Qasimi, al-Qasimi, and al-Azm, *Dictionary of Damascene Crafts (Qamus Al-Sina'at Al-Shamiyya)*. 52. Author's translation.

[17]Abu Salim, *Crafts Corporations in Damascus 1700–1750 (al Asnaf w al Tawaif al Herafyyieh fi Madenit Dimashq 1700–1750)* 259.

Table 8.1 Building crafts as described in the Dictionary of Damascene Crafts of al Qasimi. Light grey hatched crafts are both mentioned in the Builders craft (dark grey) and as an independent craft

Name of Trade-Craft	Name in Arabic		Type of Trade
Lime burner	أتوني	Atuni	Supplying
Clay worker	تراب	Tarrab	Supplying
Builder	بناء- معماري	Banna'- Mi'mari	Building
Plaster worker	جباسيني	Jabasini	Supplying
Stone extractor	حجار	Hajjar	Supplying
Rammed earth builder	دكاك	Dakkak	Building
Painter-decorator	دهان	Dahhan	Building
Wall washing worker	حوار	Hauwar	Building
Brick worker	طواب	Tauwab	Building
Mortar worker	طيان	Tayyan	Building
Mortar mixer	مجارفي	Majarfi	Building
Unskilled labour	فاعل	Fa'el	Building
Plumper	قساطلي	Qasatli	Building
Surveyor	مساح	Massah	Supplying
Engineer	مهندس	Muhandess	Supplying

initiation and they have no relation with Sheikh al-Mashaiekh. They appoint their own masters as Sheikhs and make their own regulations … a new president is appointed every three months. I was given to the effect that …they used to be initiated but later escaped from the authority of sheikh al Maskhaiekh.'[18]

Based on the two texts, it could be argued that the guild of builders, unlike any other guild, had a specific relationship with the state. This specificity is evident in court records in Damascus that do not call the head of this guild by *al-shaykh* as in other guilds, but they name this position as *mimar bashi*.[19] This position was not inherited within families, and the *mimar bashi* was not a member of the guild himself, he was placed by the orders from the government.[20] Court records also show that there were two other important positions in the builders' guild. The guild master builder (*mu'allim*) who helped the *mimar bashi,* and masters from outside the guild *Kalfa*, a word from the Arabic origin *Khailfa* that means vicar. Although it

---

[18]Qoudsi, *Notice sure les Corporations de Damas (Nubdha tarikhiyya fil hiraf al Dimashqiyya)* 59.

[19]Rafeq, 'Aspects of Crafts Organisation in Bilad Al Sham during the Ottoman Era (Mathaher Men Al Tantheem Al Herafi Fi Bilad Al Sham Fi Al Ahd Al Othmani)'.

[20]Abu Salim, *Crafts Corprotations in Damascus 1700–1750 (al Asnaf w al Tawaif al Herafyyieh fi Madenit Dimashq 1700–1750).*

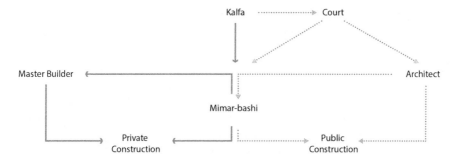

Fig. 8.2 Guild of builders construction model

is not explicitly mentioned in the case of Damascus, the term *Kalfa* in Northern Ottoman cities refers to moving artisans who wander around the empire for public and private commissions.[21]

The *mimar bashi* was not just a master builder; but neither was he a state architect. Ottoman state Architects were trained in the facilities of the army.[22] A look at the structure of the guild of builders unfolds the relationship between the two. On the one hand, the administration in the guild, headed by the *mimar bashi*, facilitated private construction and housing rehabilitation projects in a non-centralised fashion—architects were not needed, and urban planning per se was an in-situ procedure. On the other hand, through this same organisational system the *mimar bashi* recruited and contracted master builders for governmental and public projects led by state architects. The combination of governmental and popular collaboration could accommodate public and private construction by using the same builders, crafts and materials.[23]

Thus, we are facing two roles in building trade, namely the architect and the master builder; the first as synthesisers and visionaries, and the second as problem solvers. The first is related to governmental and public projects, and the second to residential and public buildings. The training in the first is institutionalised; whilst in the second is based on craftsmanship values. The guild is the place where these two roles meet (Fig. 8.2).

---

[21]Cerasi, 'Late-Ottoman Architects and Master Builders'.
[22]Cerasi.
[23]Abu Salim, *Crafts Corporations in Damascus 1700–1750 (al Asnaf w al Tawaif al Herafyyieh fi Madenit Dimashq 1700–1750)*.

8.3 The End of the Guild: Transformations in the Role of Master Builders

Thirty years after Qoudsi's essay in Liede, the guild of builders in Syria ceased to function.[24] The two roles of master builders and state architects were unified in the person of the modern architect—architecture is then a recognised profession with a different line of education than that of guild training or military establishments. Planning was shifted from the hands of the builders to the table of legislative entities. However, the end of guilds was not a sudden systematic giving up to machines. There was an important transitional period of adaptation and experimentation.[25] What is considered an acute change to Westernisation in architecture and urban planning in the last twenty years of the Ottoman rule in Damascus is an example of continuity in building construction. Western Architects, with their institutional education and 'modern' references had to work with local builders and materials which led to introducing or inventing new and different techniques such as roofing systems or façade stonework.[26] However, things were very different on popular construction level; it is in this period that the first dichotomy cracked between artisans and the city. Without the guild of builders, no institution could channel, agent or even acknowledge private construction. The term unplanned surfaced in planning— the French planner René Danger noted some of 'spontaneously built' areas and responded to them by creating 'non-aedificandi' zone.[27] Since then, the official representation to this type of private construction was by their 'illegality'. Building craftsmen were no longer seen as problem solvers but possible threats.

The second dichotomy came later, and it was by prohibiting the use of traditional materials for new construction in Syria. Before the use of concrete, unplanned areas around the wall of Damascus were integrated parts of the old city. The same materials and structural language were used in restoration, popular and architect-led projects. Until only recently, scholars categorised those unplanned areas, the *meidan*, for example, as part of the Ottoman expansion plans.[28]

The cement construction started at the beginning of the twentieth century. In 1929, Damascus inaugurated the first cement factory in its outskirts.[29] The state played a major role in advocating reinforced concrete as the only building material. In 1925, buildings were restricted to stone and bricks for safety and hygiene

---

[24]Daghman, 'Mud Building Architecture in Damascus Region, Analysis and Documentation Study (Amaret Al Abnyeh al Tinyeh fe Iqleem Dimashq- Diraset Tauthiqiyye Tahliliyye)' 21.

[25]Hanna, *Artisan Entrepreneurs in Cairo and Early-Modern Capitalism (1600–1800)* 154–188.

[26]Daghman, 'Mud Building Architecture in Damascus Region, Analysis and Dumentation Study (Amaret Al Abnyeh al Tinyeh fe Iqleem Dimashq- Diraset Tauthiqiyye Tahliliyye)' 44.

[27]Etienne, 'Mukhalafat in Damascus: The Form of an Informal Settlement'.

[28]Etienne.

[29]The cement factory was also a benchmark in the syndical movement in Syria. Workers in the factory were pioneers in forming a modern syndicate to claim their rights.

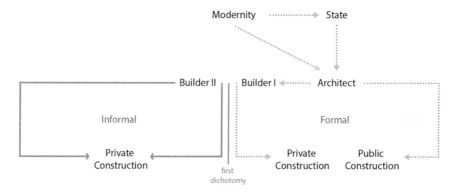

Fig. 8.3 Transitional period|construction model

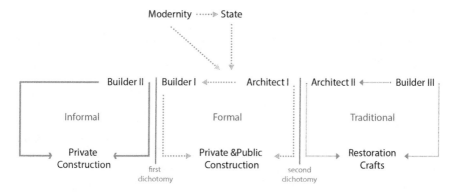

Fig. 8.4 Current construction model

reasons.[30] Reinforced concrete continued to be the building material during and after the French mandate. In 1976, wood and mud bricks were prohibited in Damascus region. The use of traditional materials gradually ceased in Damascus.[31] It was the formal construction that changed the material in the informal construction. Construction workers started to use reinforced concrete for its availability, and traditional crafts with traditional materials took its final refuge in restoration projects in old cities. Figures 8.3 and 8.4 present schematic construction models from the transitional period to the current practice.

---

[30]Daghman, 'Mud Building Architecture in Damascus Region, Analysis and Dumentation Study (Amaret Al Abnyeh al Tinyeh fe Iqleem Dimashq- Diraset Tauthiqiyye Tahliliyye)' 51.

[31]Daghman 59.

8.4 Building Crafts and Trades in Today's Construction

The result of the two dichotomies of materials and representation constitutes the modern construction is Syrian cities. Craftsmen with traditional methods and techniques, such as masonry and brickwork, work on restoration projects. Their work is marginal, isolated and mummified.[32] And the quality of their crafts is deteriorating. Crafts in general are seen as tourists' leisure.[33] The case of construction crafts, in particular, is less promising, and recent calls to revive and document craftsmanship did not include any of the construction-related crafts or was only concerned about decorative ones.[34]

But there is another extension to the traditional builders' remit beyond restoration: the building of informal settlements, the model of the mediator in a complex social context, informal agreements of privacy, and accessibility rights for its residence. This model follows structural and design regulations for maximum spans, rooms arrangement and natural light access. Popular builders advise families on how to divide lands to build in their farmlands. They are also consulted when an already built area is being densified and houses are being divided or rearticulated. Popular builders use ordinary materials available in the market. When the formal sector of construction was mainly building with cement, suppliers of traditional materials stopped trading and cement and steel prevailed in both formal and informal construction.

The difference between the modern builder and the builders in the guild is in the agency. Unlike the guild builders, present-day ones have no formal representation with the centralised authorities. The relationship between the two is not pronounced and involves bribes and connections.[35] The construction of a house in an informal settlement is based on temporarily blinded-eye methods, where authorities give only a few days to the builder to finish the main structure of the house, i.e. the roof and walls. Speed in construction is now one of, if not the most important skill a builder should have.[36]

What once was a regulated contest between the state and the builder, is now an open one. Planners in Damascus, since Rene Danger, treat informal settlements as large already built areas and 'urban facts' that pops up out of the blue. Planning happens on massive scales in informal settlements, with micro interference that recognises the popular builder is limited.

---

[32]Abed, 'Traditional Building Trades and Crafts in Changing Socio-Economic Realities and Present Aesthetic Values' 73.

[33]The official governmental sponsor and supporter of craft in Syria is the Ministry of Tourism. See The Syria Time, 'Tourism Ministry'.

[34]See the list of the intangible cultural heritage in Syria by the Syrian Trust for development http://ich-syr.org/.

[35]Etienne, 'Mukhalafat in Damascus: The Form of an Informal Settlement'.

[36]Etienne.

8.5 Discussion: Artisans and Master Builders to Rebuild Syrian Cities

Further research in guild literature is open for architects and planners. In this paper, the model of the guild of the builder in the late Ottoman empire in Damascus is drawn up through focusing on two texts from that period and crossing them with guild research on Ottoman courts documents. The model is a speculative proposal that needs further verification. However, what is interesting about the Ottoman guild of builders is that, despite its authoritarian structure as a guild, it is inclusive to different types of building practice, it deals with planning conflicts of neighbours —with other neighbours and the state by both in situ applications and courts orders. It is also inclusive to both private and public construction, local and foreign building techniques.

Since 2011, the conflict in Syria resulted in an extreme destruction of cities, in historical and modern centres, rural and urban areas. The unfortunate destruction of the traditional building has brought with it a possibility to re-link local materials to popular building practice. The restoration of the vaulted souk in Aleppo, for example, has raised a possibility to learn about and use vaults in new housing projects. The question of how those two processes could be interlaced is an extremely valid question. How could the traditional crafts used in restoration projects be effective elsewhere?

Despite all the efforts to control the construction in Damascus, around 40% of its built environment is built by popular builders with minimum, if any, intervention from architects or planners. Architects are not the primary key players in this production and, in some cases, are completely absent. During the conflict, the illegal construction in those areas almost doubled, taking advantage of the absence of authorities.

8.5.1 The Master Builder

By looking at building practice of the guild structures and how it was dissolved during modernisation, informal settlements refrain from being a phenomenon or an urban fact that comes with modernity. Instead, it becomes a continuation of the private construction sector in the guild but in a hostile context of laws and material market.

It is impossible to have an exact reproduction of guilds in today's modern forms of governance; a recovery of the master builder is nonetheless still possible. The reconstruction of the destroyed areas in Syria has to open up to informal building experience. In such a delicate phase, there is a genuine need in not only planning nor in brute force labour but also in the multidisciplinarity of labour that can respond to different social and technical context.

8.5.2 Training

Popular builders are no longer considered as artisans, but they still have apprentices and journeymen in an informal training that challenges the institutional vocational education provided by the state. The difference between the skilled and the unskilled labour in such a context is not about the work of the hand and the perfection of the work, the unskilled informal builder becomes skilled only when he or she can respond and take decisions about complex and changing challenges such as dividing lands and expanding houses. It is also true that such a builder can only be independent when he can secure a specific set of techniques to avoid or accommodate the obstacles posed by planning and state authorities.

8.5.3 Knowledge Exchange

Another important role to recover from the guild builder is that of Kalfa, the moving artisans. Syria is neighbours with areas and cultures with similar rich building traditions. A research into what construction crafts could be transferred from those cultures is very much needed and is the extension of this paper. Technology transfer had an important role not only on the style and trends of architecture in the Middle East but also in developing new structural vocabularies and construction habits. It is by the time materials were transferred instead of knowledge that the building practice become generic, standardised, causing amnesia that raises important questions about identities Middle Eastern Cities.

Returning to building crafts and local materials in times of crisis is not new. During the sanctions on Syria in the 1980s, a severe shortage of steel in the market drove the construction market to find alternative solutions. Builders and architects in the formal sector looked at other types of roof structures on load bearing walls and vaults were adopted. The Military Housing Corporation representing a state construction organisation, built several vaulted housing projects in Damascus and Aleppo.[37] There were no such craftsmen as vault makers in Syria, but the Military Housing Corporation worked as a 'master builder' by conducting experiments, tests and samples of brick-vaulted house. Today's technology in structure and materials fabrication is more master builder-friendly. Systems of knowledge sharing, training and manufacturing are also rapidly advancing.

The artisan work on construction and its future role in rebuilding Syria is, therefore, about the holistic role of the master builder. This path can be taken not by training labour to be skilled only, but also to train the architect to be skilled builders.

---

[37]Abed, 'Traditional Building Trades and Crafts in Changing Socio-Economic Realities and Present Aesthetic Values' 36.

References

Abed JH (1988) Traditional building trades and crafts in changing socio-economic realities f and present aesthetic values: case studies in Syria. Thesis, Massachusetts Institute of Technology. http://dspace.mit.edu/handle/1721.1/62895

Abu S, Isa SM (2000) Crafts corprotations in damascus 1700–1750 (al Asnaf w al Tawaif al Herafyyieh fi Madenit Dimashq 1700-1750). Al Fikr, Amman

Arnakis GG (1953) Futuwwa traditions in the ottoman empire akhis, bektashi dervishes, and craftsmen. J Near Eastern Stud 12(4):232–247

Baer G (1970) Guilds in Middle Eastern history. In: Cook MA (ed) Studies in the economic history of the middle east 1970. Psychology Press, pp 11–30

Cerasi M (1988) Late-Ottoman architects and master builders. Muqarnas 5:87–102. https://doi.org/10.2307/1523112

Daghman M (1994) Mud building architecture in damascus region, analysis and documentation study (Amaret Al Abnyeh al Tinyeh fe Iqleem Dimashq- Diraset Tauthiqiyye Tahliliyye)

Etienne L (2012) Mukhalafat in Damascus: the form of an informal settlement. In: Ababsa M, Dupret B, Dennis E (eds) Popular housing and urban land tenure in the middle east: case studies from Egypt, Syria, Jordan, Lebanon, and Turkey. http://cairo.universitypressscholarship.com/view/10.5743/cairo/9789774165405.001.0001/upso-9789774165405-chapter-2

Hanna N (2011) Artisan entrepreneurs in Cairo and early-modern capitalism (1600–1800). Syracuse University Press. http://www.jstor.org/stable/j.ctt1j1w0jp

Khaldūn I (1969) The Muqaddimah : an introduction to history, 3 vols. 1. Princeton University Press

Kuban D (2001) The miracle of divrigi. Yapi Kredi, Istanbul

Lewis B (1937) The Islamic guilds. Econ History Rev 8(1):20–37. https://doi.org/10.2307/2590356

Massignon L (1953) La Structure Du Travail À Damas En 1927: Type d'Enquête Sociographique. Cahiers Internationaux de Sociologie 15:34–52

Meinecke M (1985) Mamluk architecture, regional architectural tradition: evolutions and interrelations. Damaszener Mitteilungen 2:163–175

Qasimi MS, Jamal al DQ, Khalil al-Azm (1960) Dictionary of damascene crafts (Qamus Al-Sinaʿat Al-Shamiyya). In: al-Qasimi Z (ed). Mouton, The Hague

Qoudsi E (1992) Notice sure les Corporations de Damas (Nubdha tarikhiyya fil hiraf al Dimashqiyya). Al Hamra, Beirut

Rafeq A-K (1981) Aspects of Crafts Orgenisation in Bilad Al Sham during the Ottoman Era (Mathaher Men Al Tantheem Al Herafi Fi Bilad Al Sham Fi Al Ahd Al Othmani). History Stud (Dirasat Tareekhyy) 4:30–62

Rafeq A-K (1991) Craft organization, work ethics, and the strains of change in Ottoman Syria. J Am Oriental Soc 111(3):495–511. https://doi.org/10.2307/604267

The Syria Time (2013) Tourism ministry: cooperation with Civil Society Organizations Stressed. Syriatimes.sy. 22 Sept 2013. http://syriatimes.sy/index.php/tourism/8349-tourism-ministry-cooperation-with-civil-society-organizations-stressed

Wolper ES (2003) Cities and Saints: Sufism and the Transformation of Urban Space in Medieval Anatolia. Pennsylvania State University Press

Chapter 9
Place-Identity in Historic Cities; The Case of Post-war Urban Reconstruction in Erbil, Iraq

Avar Almukhtar

Abstract Throughout history, war has caused fundamental political, economic and social transformations around the world, spatially impacting urban form. This is evident in cities undertaking post-war reconstruction where global influences on unique historical landscapes and the cities' distinctive identities can be observed. Erbil, the capital of the Kurdish region of Iraq is such a city. The city hosts the Erbil Citadel, a UNESCO World Heritage Site, which dates back to nearly 5000 B.C. After war ravaged the country in 2003, a decade of reconstruction ensued with a long period of political and economic stability resulting in rapid urbanisation. The post-war reconstruction process was a challenge between aspirations to promote the city globally (as the capital of an emerging nation) and a desire to represent Erbil's historical roots in the Citadel and the old town. This has involved an array of urban actors including international investors and NGOs, who have influenced the transformation of Erbil's place-identity. This chapter studies the transformation of Erbil and its historical identity during the post-war reconstruction period (2003–present). It employs a morphological analysis of the city's historic core and contemporary areas combined with knowledgeable interviews with key policy makers, locals and stakeholders in order to explore the impact of the intensive post-war urban development process on the city's place-identity. Key findings of Erbil's morphological analysis indicate that the post-war reconstruction process has radically transformed the city's urban fabric both in the old and the contemporary areas. Arguably, this has reflected globalised design patterns and ignored the historic morphological traces that the city has acquired throughout centuries. Consequently, post-war urban transformation has been negatively impacting Erbil's unique place-identity. There is an urgent need for a comprehensive urban design approach that can guide future development in the contemporary historic city of Erbil to enhance the continuity of its urban cultural heritage. More importantly, an approach that allows for the evolution of place-identity needs to be rooted in its rich historical and cultural values while looking forward to future opportunities.

A. Almukhtar (✉)
Oxford Brookes University, Oxford, UK
e-mail: a.almukhtar@brookes.ac.uk

Keywords Place-identity · Heritage · Erbil · Urban design · Post-war · Reconstruction process · Urban transformation · Recovery

9.1 Introduction

Most historic towns and cities have their own character and identity, which distinguishes one place from another. Place-identity is related to the quality of a place being unique and rooted in its local context (Southworth and Ruggeri 2011). Although the concept of place-identity is engrained in history, it is neither static nor uniform (Sepe 2013; Watson and Bentley 2007). It is the result of a continuous evolutionary process through the interaction of humans and place during different cycles of civilisation (Sepe 2013; Magnaghi 2005). This process is constantly influenced by the dynamic changes of social, political and economic forces over time (Lynch 1981). This is argued by Carta when he says that place-identity *'is not a static image of its state, but is rather the result of concrete development over time. This is due to the fact that identity is the outcome of relationships established between people and their environments. By making their mark on a region's cultural heritage'* (Carta 1999, p. 151 cited in Sepe and Pitt 2014). One of the crucial factors that shapes place-identity is heritage as it represents the values and traditions of the past within the present, connecting people's past with their present and future while reiterating the sense of national and regional identity (Ashworth and Larkham 2013; Graham 2002). Hence, it is strongly linked with nation-building and national identity. For example, the Acropolis in Athens is a symbol of Greece and the national identity of Athenians (Rakic and Chambers 2008).

The city of Erbil is the capital of the Autonomous Kurdish Region of Iraq. The city is going through an intensive post-war reconstruction process which is impacting its place-identity. Historically, many nations have consisted of different ethnic and cultural groups sharing the same geographical place. Some of these ethnic groups are considered minorities despite their distinctive culture, history and local identity (Oliver 1997). Often their culture and identity is neglected, changed, or destroyed when different political ideologies practiced by the dominant ethnic group exist, thus creating conflict. However, this attack on identity is not only on the human level, it also includes an assault on traditions and historical values, heritage sites and architecture (Piquard and Swenarton 2011). Once the situation stabilises, reconstruction begins as part of the post-conflict recovery process (Minervini 2002). This includes political, social and economic changes, which in many cases result in the transformation of character and identity of the place. Often less responsive to the local context, this evolving place-identity may threaten the local urban fabric and heritage. The Kurdish minority and the city of Erbil in Iraq are clear examples of such a situation and are the focus of this research where post-war urban transformation has been negatively impacting Erbil's place-identity.

9.2 Research Methodology

This book chapter is based on the Ph.D. research conducted by the author. The study analyses the historical transformation of place-identity in Erbil, to examine the impact of the post-war urban reconstruction process on the city's place-identity. To examine this process, the research used in-depth morphological analysis (Oliveira 2016) on-site in Erbil, supported by discourse analysis of 47 semi-structured interviews with local residents and various key stakeholders, who make decisions concerning the development of the built environment and its transformation process. The interviews included government officials such as the Director of the Urban Planning Department, the Deputy Head of the High Commission for Erbil Citadel Revitalisation, architects from local municipalities from each of the case study areas, and urban designers and planners from the Ministry of Municipalities and the governorate of Erbil. In addition, interviews were conducted with professionals such as architects and urban designers working in the private sector, university academics and developers. Interviews sought to explore their perceptions of place-identity, and their involvement in the development process. Additionally, the morphological analysis included a review of historical maps, current and historic master plans, photos and reviews of planning policies and documents in order to understand how the political conflict influenced the planning and urban design mechanisms and consequently transformed the place-identity through time within Erbil's urban fabric.

9.3 Context

While a deep political understanding is important to explain the Kurdish conflict, this is not the focus of this research. However, various political facts will be used to discuss the Kurdish context as a foreground for the spatial consequences of the post-war reconstruction.

Kurds are the world's largest ethnic group without a state who share common and distinct culture, traditions, language and identity (Gunter 2008; Aziz 2011). The geographic area they inhabit, Kurdistan, is divided between four countries; Iran, Turkey, Syria and Iraq. However, it does not have an official boundary and it is not an internationally recognised state (Stansfield 2003; Diener and Hagen 2010). In Iraq, this area is also shared with different ethnicities such as Arabs and Turkmens, with Arabs being the politically dominant majority. Iraq was established as a Kingdom in 1920, a period where Kurds were involved in political and military conflict with the Arab government to protect their right to practice their culture and maintain their identity (Stansfield 2003; Tripp 2007). When the country became a republic in 1958, the national government's provisional constitution recognised Kurds as equal citizens for the first time (Chaliand 1994; Anderson and Stansfield 2005). This national government was later overthrown by a Baa'thist coup in 1963

(Tucker and Roberts 2010) and introduced Pan-Arabism (Arnold 2008), forcing all non-Arab ethnicities including Kurds to comply with the Arab way of life and culture (Manafy 2005; Stansfield 2007). These Arabist ideologies and policies resulted in nearly five decades of armed conflict, genocide, mass executions and human rights violations towards the Kurds. The city of Erbil, the focus of this research, is a source of identity and pride amongst the Kurdish nation. In addition to the physical fabric of the Citadel, a world heritage site, the city has a symbolic link to the immaterial heritage of the Kurds. Moreover, Erbil has gone through different periods of governance, as well as armed and political conflict and represents the capital of an emerging nation (Almukhtar 2016).

9.4 Historical Development of Erbil (Morphological Analysis of Erbil)

Erbil is the capital of the Kurdish region of Iraq and is located in the north of the country. The city is approximately 6000 years old, having originated from a surviving ancient settlement called the Citadel. The Citadel is thought to be the oldest continuously inhabited settlement in the world and is built on top of an artificial mound raised up by the process of building and rebuilding of structures in a form that has evolved naturally over thousands of years (Yaqoobi et al. 2012). It represents layers of multiple civilisations from the Neo-Sumerian times when the first courtyard housing typology appeared until the end of the Ottoman Empire in 1918. Therefore, the urban fabric was subject to various planning laws and regulations ranging from the organic development of the urban fabric based on people's socio-economic needs to the adoption of Islamic planning principles. However, most of the existing urban structure of the Citadel relates to the period of the Ottoman Empire (Yaqoobi et al. 2012) (Fig. 9.1).

The urban fabric inside the Citadel consists of a complex organic network of alleyways and open spaces connecting a hierarchy of urban spaces, transitioning from public to semi-public to private to provide privacy to the inhabitants (Fig. 9.3). Similarly, the plots in the Citadel developed in an organic pattern from the main gate in the south and then gradually covered the whole Citadel area with the majority being residential. The Citadel also includes various types of housing such as larger mansions for the rich called *Diwakhana* and *Iwans* (Yaqoobi et al. 2012), as well as smaller traditional courtyard houses. As privacy was an important element of cultural identity, traditional houses had large windows overlooking their courtyards instead of the external alleyway. Meanwhile, only small ventilation openings were placed at high levels of these walls in order to protect the privacy of the inhabitants (Fig. 9.2). Therefore, courtyard typologies were environmentally and socially responsive to the residents.

The character of the alleyways, plot pattern and typology of buildings are the result of a multitude of individual and family decisions on how each house should

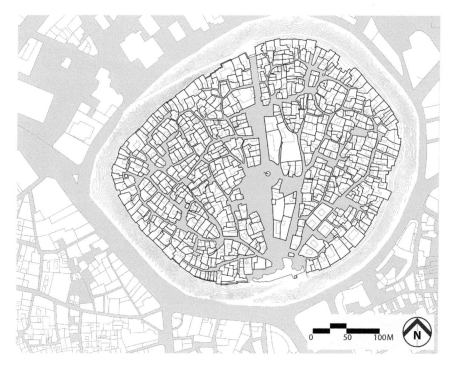

Fig. 9.1 Erbil Citadel existing urban fabric and courtyard housings. *Source* Author (2017)

look which reflect their lifestyle, financial capability and social needs. These all transformed and evolved over centuries based on the socio-economic needs and interests of residents allowing for a diverse demographic spectrum. Therefore, the network of alleyways, plots and buildings evolved organically, the exact opposite to the planned approach in which everything is pre-determined by planners, architects or urban designers before it is built.

Although the city started to slowly grow around the mound during the 12th century, the Citadel still remained the heart of Erbil, whose rich and varied history contributes to the formation of its distinctive identity (Almukhtar 2016; Yaqoobi et al. 2012). In the interviews undertaken with residents by the author, many indicated that the Citadel is a source of pride and provides a sense of belonging. As one elderly man said, '*I feel nostalgic about the Citadel because my parents and grandparents were born there. I feel my roots are from the Citadel and that I belong here. It makes me feel proud to live in Erbil and be Erbili*'. It tells the story of how hundreds of past generations interacted with their natural environment and how they developed a way of life based on their socio-cultural values and needs. The Citadel now stands as a symbol of Kurdish history, identity and culture within civil society and the government (High Commission of Erbil Citadel Revitalization 2016). As mentioned in the World Heritage nomination dossier by United Nations

Fig. 9.2 Erbil Citadel courtyard housing. *Source* Author (2015)

Fig. 9.3 Erbil Citadel housings and network of alleyways. *Source* Author (2017)

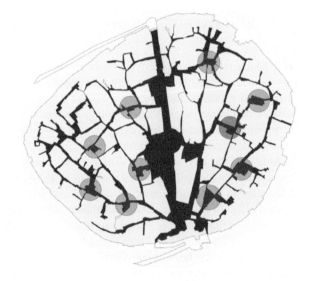

Educational, Scientific and Cultural Organisation: *'The Citadel of Erbil is a rare surviving example of an urban ancient settlement which developed on an archaeological tell, following layer by layer and time after time, a spontaneous, nonplanned growth that was influenced by a combination of previous urban layouts and successive architectural and urban elements, in a continuous process of*

addition and transformation extending back at least 6000 years, to the earliest phase of urbanism.' (UNESCO 2014).

The initial development of the old town marks the first expansion of Erbil outside the Citadel to the lower plain and is the area around the Citadel mound, which developed in the twelfth century near the southern main gate (Fig. 9.4). The development of the lower town started with the main bazaar and then grew to include a mosque, a cemetery, a school and residential structures (High

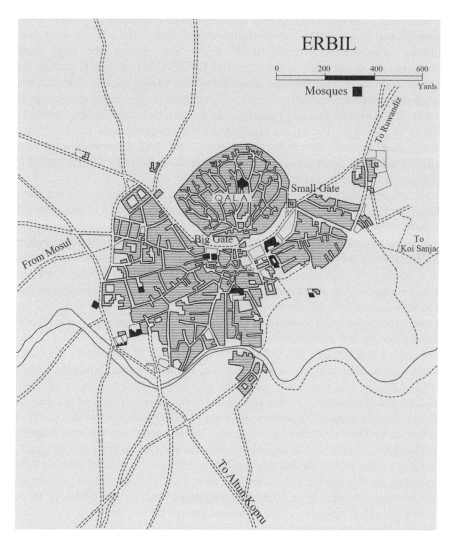

Fig. 9.4 Erbil map of 1944. The first expansion of the city outside the Citadel towards the South where the main gate is located. *Source* Drawn by the author based on a map from Ministry of Municipalities, KRG

Commission of Erbil Citadel Revitalization 2016) and it represents a unique semi-organic urban fabric that grew radially from its centre.

The morphological analysis of the old town is characterised by a pattern of narrow, irregular alleyways and streets with cul-de-sacs, most of which did not allow vehicular access. Similar to the Citadel, this network was either planned by the residents of the surrounding dwellings to provide access or emerged over time as a result of incremental growth. Historically, with the absence of a planning system, plot patterns and building typologies were based on decisions by individuals and families whose private socio-economic interests transformed the typology of residential plots and buildings over time. The range of different plot and building sizes resulted in a socially diverse neighbourhood with various family sizes from different socio-economic backgrounds living and sharing the same area. Furthermore, the old town is considered to be the oldest settlement outside the Citadel with the largest heritage urban fabric. Its morphological character was developed following similar typologies which existed in the Citadel (Ibrahim et al. 2014). The prevalent building typology was the historic courtyard brick house of one or two storeys. The layout of the residential dwellings was influenced and characterised by many factors, such as climatic conditions, the availability of local building materials, construction methods and social values. However, this started to change towards the mid-twentieth century.

Although the city later developed radially beyond the Citadel and the old town, the urban fabric started to change dramatically towards the end of the twentieth century. This was due to different political ideologies, design standards and the application of national planning policies that had an impact on the economy, culture and lifestyles of the local people inhabiting the area. With the establishment of Iraq as a country in 1920 under a British mandate, the planning system started to develop based mostly on British town planning principles (Nooraddin 2004). This was followed in 1958 by the establishment of the republic of Iraq when the first architecture school was established in Baghdad. The national government was overthrown by the aforementioned Baa'thist coup in 1963 (Nooraddin 2012). Decades of Baa'th party rule influenced many aspects of the country, including planning and urban design policies, processes, and architecture. While the historic urban fabric in Iraq was the result of the evolution of different layers of ethnic cultures and civilizations, Arabic and Islamic towns were considered the national architectural heritage of the country. All town planning and development strategies as well as building regulations were decided by the central authorities in Baghdad and applied elsewhere in the country ignoring the culture, tradition and heritage of minority ethnic groups (Nooraddin 2004). For example, the master plan for Erbil was prepared in Baghdad with little input from local officials and no civil society involvement (Bornberg et al. 2006). It represented a modular grid system of streets cutting through the existing urban fabric and ignored the original circular growth of the city (Nooraddin 2012).

Consequently, areas which were developed beyond the Citadel and the lower old town in Erbil contrasted with the traditional urban fabric that the city had acquired for centuries (Fig. 9.5). However, the city maintained its radial growth pattern. This

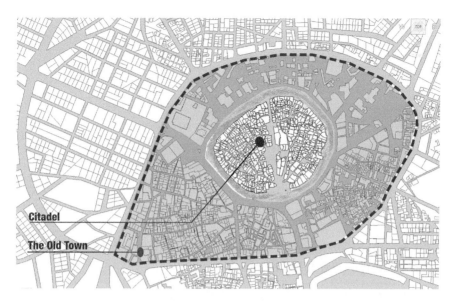

Fig. 9.5 Development of the old town around the Citadel. *Source* Author (2017)

has also resulted in the transformation of plot typologies from irregular semi-organic shapes with a diverse range of sizes to regular geometric shapes with a limited variety of plot sizes targeting particular demographic groups based on their social and financial level. This was obvious in the development of various new neighbourhoods such as the 'Engineering neighbourhood' and the 'Teachers neighbourhood' which resulted in social divisions. Meanwhile, the building typologies marked a departure from traditional brick courtyard houses which reflected socio-cultural needs as opposed to Western-style houses without courtyards and the use of imported building materials and techniques. Hence, the local place-identity of Erbil started to change gradually as one moved out of the Citadel and beyond the city centre. Consequently, this has resulted in inadequate urban planning policies and frameworks that do not respond to local socio-cultural needs and priorities. Therefore, applying Arab national policies and development patterns in the region attempted to manipulate and change the rich heritage asset and local place-identity of the area. The heritage sites and traditional urban fabric of old towns and cities were demolished or replaced in the Citadels in Kirkuk and Erbil and other parts of the Kurdish area (Nooraddin 2012). The demolition in the 1970s of the Citadel's historic gate by the Ba'th government is a clear example. It was replaced with one that included architectural elements imported from Babylonian design from outside the region and was felt by many to be an attempt to 'Babylonise' the Citadel (Fig. 9.6) (Nooraddin 2012).

Fig. 9.6 Citadel gates. (left) The recently Babylonian style gate, (right) The redesigned gate based on the original. *Source* Author (2012, 2016)

9.5 Post-war Urban Reconstruction; A Challenge Between Heritage and Globalisation

As outlined above, decades of political instability starting at the beginning of the last century had consequences on the urban landscape of the city in which heritage sites experienced neglect and lack of conservation. This was followed by a period of economic growth after 2003 that led to rapid post-war reconstruction as part of the recovery process. The extraction of oil from Kurdish-dominated areas has improved the economy of the region. Political stabilisation has resulted in strong growth in the private sector, with local and international investment opportunities. The Kurdish government's efforts to develop and upgrade the region's infrastructure have transformed the urban environment (Almukhtar 2016). Erbil gained political significance as the capital of Iraqi Kurdistan in 1991 when the region became autonomous. Economic growth from oil revenues and United Nations support for rebuilding the area have transformed the city into a hub for investment, trade, tourism and development in the region (Almukhtar 2016).

The Kurdish government's first steps were to restore and preserve the Citadel as it represents the heart of Erbil, its identity and sense of belonging (Bornberg et al. 2006). Consequently, the significance and the uniqueness of its distinctive historic urban fabric and vernacular qualities focused efforts to have it included in UNESCO's list of World Heritage Sites in 2014 (UNESCO 2014). However, in an attempt to address political decisions adopted by the Baa'th party during its control over the city that affected urban heritage and identity, the Babylonian-style gate was demolished. Consequently, a decision was made by the Kurdish government to demolish the reconstructed gate and replace it with one that followed the original historic design (Fig. 9.6). However, they did not consider that the 1970s Babylonian gate was a part of the Citadel's evolution process; nor did they view it as a legitimate part of the city's transformed identity, highlighting the evolving nature of place-identity in historic urban settings and areas affected by conflict.

Although places with built heritage have the potential to provide cities with a distinctive place-identity through rehabilitation and conservation programmes, this was a challenge in the case of Erbil as heritage sites are marketed to promote economy, political agendas and tourism. Also, the inclusion of the Citadel in the World Heritage Sites List was challenging. Once a place receives official recognition as a heritage site, its relationship with the its landscape and people changes. On the one hand, the outstanding universal value is a tremendous recognition as an asset to the city's character and identity. On the other hand, it can create limitations for development and the natural evolution of a place in the context in which it exists (Governorate of Erbil 2013; Jansson 2010). In Erbil, it is very challenging to accommodate the need for modernisation and post-war reconstruction interventions without compromising the city's character and identity. Additionally, managing the competing interests of different stakeholders with the potential for tourism and economic growth might negatively impact the local place-identity if it is not guided towards simultaneously respecting the inherited heritage and its landscape setting (Labadi and Bandarin 2007). This post-war reconstruction was an opportunity for the Kurdish government to use a culturally heritage site in the post-war development to represent an emerging nation that is globally associated with its rich heritage, also, to promote the city and Kurds using the heritage. However, reflecting and respecting this asset within the development process outside the Citadel has arguably failed.

This desire to compete on a regional and global level combined with an economic boom and political stability has led to a strong drive for rapid post-war reconstruction as part of the recovery process (Almukhtar 2016). All this has resulted in rapid urban transformation across the region and particularly in Erbil, changing the city's local place-identity through its evolving urban fabric and new developments. The morphological analysis showed that the reconstruction process outside the Citadel was mostly globally influenced with less consideration given to the local urban fabric and the cultural values associated with it. A clear example is the new housing typologies which represent a significant theme of globally influenced development that has covered the majority of the post-war urban developments.

The limited urban development in Erbil, which is due to decades of war and severe sanctions, together with the recent influx of Internally Displaced People (IDP) and refugees have led to a shortage of housing across the city and an escalation in rental prices (Almukhtar 2016). Consequently, housing policy and development became one of the first steps taken towards the post-war recovery process, although the process of housing production in Erbil mostly involved local people building houses for themselves based on their financial capabilities and socio-cultural needs and preferences either through architects or master builders. The land in most of the residential areas was owned by the state and was granted either for free or at a low price for government employees (KRG, Ministry of Municipalities and Tourism 2013). However, in the post-war reconstruction process another approach for housing development was introduced in which the Kurdish government introduced new initiatives where investors could purchase large areas

Fig. 9.7 Gated communities segregated from the surrounding context both socially and spatially. *Source* Author adapted from Nawzad Ali (2016)

of land with symbolic prices to produce residential complexes. The Kurdish government also introduced affordable housing schemes that offered mortgage loans targeting low-income groups, and this has led investors to develop cost-effective housing with limited consideration to socio-cultural needs and local place-identity. Moreover, the rapid economic growth has attracted both local and international investors (Kurdistan Board of Investment 2016). They have developed housing projects that reflect global design ideologies in order to target high-income segments of society. Examples of these types of developments are the 'Italian Village' and the 'English Village' (Fig. 9.8), where English and Italian terms are used for marketing purposes and they also reflect, to a limited extent, architectural styles implied by the names of these developments. Similarly, the urban fabric and design typologies used lack responsiveness to the local culture and climate and are isolated from the surrounding context as they introduce western identities that are different from the local built form and pattern of the area (Fig. 9.7). Consequently, these residential developments have resulted in social divisions and they have mostly failed to provide locals with their basic housing requirements.

Both types of housing mentioned above are considered gated developments with closed perimeter walls and strictly controlled entrances that target a specific demographic of residents. Furthermore, most of these gated communities are composed of hundreds of multiple units, which are modular in type, and which lack variety and distinctiveness. As a result, these housing typologies contrast with Erbil's residential landscape in terms of their forms and functions and they have resulted in social segregation between residents of the city. Additionally, during this period, the Kurdish government introduced new investment policies and incentives in order to attract and encourage foreign investors and developers to take part in the building and reconstruction process of modern Kurdistan (Kurdistan Board of Investment 2016). However, this resulted in developments following new design ideologies that are globally influenced and architectural elements that lack

Fig. 9.8 Italian village (on the top), English village (on the bottom). *Source* Nawzad Ali (2016)

consistency with the culture, climate and the existing context of the region. The new developments reflect global influences and international architectural norms that could belong to any part of the world, and which contrast with the traditional values of the city and which lack historical continuity, thus ignoring the local place-identity of the area (Figs. 9.7 and 9.8).

However, due to its weaknesses, the current planning system does not regulate residential developments or require developers to seek appropriate skills and awareness leading to consideration of 'place-identity' from architects and urban designers throughout the design process. Additionally, research interviews with the urban planning department in Erbil revealed that the issue of place-identity is not taken into account when the department designs, assesses, reviews and approves schemes and developments outside the old town. Consequently, these developments lack character, distinctiveness and connectivity to the surrounding urban fabric and speak an urban language that is in stark contrast with the heritage and traditional Kurdish models of inhabitation and social interaction. Consequently, all these have resulted in post-war reconstruction development which is globally oriented and politically driven. Thus, Erbil has become a city made up of a collage of micro-identities conflicting and competing with one another. Most importantly, these micro-identities have failed to represent the city collectively, its rich heritage and culture, to connect local people with their cultural heritage and most importantly connect various social classes of the society together. Heritage sites can be used to connect past values, histories and traditions to the present and future as they are sites where people connect either physically or emotionally with a place and they are bound up in notions of belonging and identity. As Landry states, *'we are connected to our histories and our collective memories* via *our cultural heritage which fastens our sense of being and can supply us with a source of insight to be of help to face the future'* (Landry 2006: 6).

9.6 Conclusion

In this chapter, the impact of the post-war reconstruction process on place-identity was analysed using the city of Erbil as a case study. Place-identity in post-war Erbil has rapidly transformed, especially in areas outside the Citadel and the old town. This transformation has negatively impacted the local place-identity as many areas have lost links with their heritage and historical past and are embracing micro-identities, that can be found anywhere around the world. The city presents a collage of multiple identities that fail to collectively represent the city's local place-identity.

It can be argued that Erbil's sense of place is heavily based on the Citadel. Yet, while the Citadel can create opportunities for diverse groups of people to connect, Erbil's sense of place goes beyond just that area. Although the government encourages and provides opportunities for new modern developments as part of promoting Erbil as the capital of Kurdistan, its main focus and effort is on the heritage value of the Citadel and the heritage buildings in the old town. However, the emphasis on heritage is mostly focused on conservation and not on developing a holistic comprehensive approach to urban form which can contribute to the overall local place-identity of the city. Therefore, there is an urgent need for planning and urban design strategies that can guide future developments to reflect and respect heritage values, tradition and culture while allowing and embracing modern values. Urban design strategies and frameworks should consider ways to enhance place-identity throughout the entire city instead of myopically focusing only on heritage areas in isolation. Also, stakeholders such as government officials, urban designers, planners, architects, developers, investors and NGOs who are involved in the reconstruction and rebuilding the city must be particularly aware of the multiple identities that exist. In order to avoid spatially redesigning out of existing identities at neighbourhood, district, city and regional levels.

Without urban strategies targeted on place-identity, place-identity is at risk of evolving in a way that threatens Kurdish values, traditions and culture, and which fails to adapt and respond to the rapidly changing political and economic conditions. Most importantly, cities in the post-war context need urban design strategies that are resilient and which can adapt to political and economic changes in order to guide the reconstruction process without compromising local heritage, traditions, and identity. It is important for such strategies to be rooted in traditional culture, while at the same time being a part of global modernisation processes that introduce economic and political opportunities that will inevitably transform the sense of place. Hence, place-identity needs to be at the core of the development process in places with heritage assets to strengthen its cultural value and guide urban developments without compromising local traditions. This should be achieved in a way that values the present while looking at the past as resources for the future that embraces local roots in the globalised world.

References

Almukhtar A (2016) Conflict and urban displacement. Urban disaster resilience: new dimensions from international practice in the built environment. Routledge, NewYork
Anderson LD, Stansfield GRV (2005) The future of Iraq: dictatorship, democracy, or division? Palgrave Macmillan, New York, N.Y., Basingstoke
Arnold JR (2008) Saddam Hussein's Iraq. Twenty-First Century Books
Ashworth GJ, Larkham P (2013) Building a new heritage (RLE Tourism), Routledge Library Editions: Tourism. Routledge
Aziz MA (2011) The Kurds of Iraq: ethnonationalism and national identity in Iraqi Kurdistan. London, I.B., Tauris
Bornberg R, Tayfor MA, Jaimes M (2006) Traditional versus a global, international style: Aarbil, Iraq. Retrieved on October, 7, 2013
Carta M (1999) L'armatura Culturale Del Territorio. Il Patrimonio Culturale Come Matrice di Identità e Strumento Per Lo Sviluppo. FrancoAngeli, Milano, Italy
Chaliand GR (1994) The Kurdish tragedy. Zed Books in association with UNRISD, London
Diener AC, Hagen J (2010) Borderlines and borderlands: political oddities at the edge of the nation-state. Rowman & Littlefield Publishers, Lanham, MD
Governorate E (2013) Planning and building regulations for the buffer zone of Erbil Citadel
Graham B (2002) Heritage as knowledge: capital or culture? Urban Stud 39:1003–1017
Gunter MM (2008) The Kurds ascending: the evolving solution to the Kurdish problem in Iraq and Turkey. Palgrave Macmillan, Basingstoke
Ibrahim R, Mushatat S, Abdelmonem MG (2014) Authenticity, identity and sustainability in post-war Iraq: reshaping the urban form of Erbil City. J Islamic Architect 3:58–68
Investment KBO (2016) Kurdistan regional government [Online]. http://cabinet.gov.krd/p/page.aspx?l=12&s=030000&r=315&p=228&h=1. Accessed 21 Sept 2016
Jansson BG (2010) The significance of world heritage: origins, management, consequences: the future of the world heritage convention in a nordic perspective. The significance of world heritage: origins, management, consequences, Falun, December 8–10, 2010; The Future of the World Heritage Convention in a Nordic Perspective, Vasa, December 13–16, 2011, 2013. Högskolan Dalarna
KRG, Ministry of Municipalities and Tourism (2013) 'Erbil City Master Plan' report
Landry C (2006) The art of city-making. Routledge
Lynch K (1981) A theory of good city form. MIT Press, Cambridge, Mass
Labadi S, Bandarin F (2007) World heritage: challenges for the millennium. UNESCO
Magnaghi A (2005) The urban village: a charter for democracy and sustainable development in the city, Zed books

Manafy A (2005) The Kurdish political struggles in Iran, Iraq, and Turkey: a critical analysis. University Press of America

Minervini C (2002) Housing reconstruction in Kosovo. Habitat Int 26:571–590

Nooraddin H (2004) Globalization and the search for modern local architecture: learning from Baghdad. In: Elshishatawi Y (ed) Planning middle eastern cities: an urban kaleidoscope in a globalizing world. Routledge, London, pp 59–84

Nooraddin H (2012) Architectural identity in an era of change. Developing Country Stud 2:81–96

Oliveira V (2016) Urban morphology: an introduction to the study of the physical form of cities. Springer, Cham

Oliver P (1997) Encyclopedia of the vernacular architecture of the world. Cambridge University Press, Cambridge

Piquard B, Swenarton M (2011) Learning from architecture and conflict. J Architect 16:1–13

Rakic T, Chambers D (2008) World heritage: exploring the tension between the national and the 'universal'. J Heritage Tourism 2:145–155

Revitalization HCOEC (2016) High commission of Erbil Citadel revitalization [Online]. http://www.erbilcitadel.org/index.php. Accessed 01 March 2016

Sepe M (2013) Planning and place in the city: mapping place identity. New York, Routledge, Taylor & Francis Group, London

Sepe M, Pitt M (2014) The characters of place in urban design. Urban Design Int 19:215–227

Southworth M, Ruggeri D (2011) Beyond placelessness

Stansfield GRV (2003) Iraqi Kurdistan: political development and emergent democracy. RoutledgeCurzon, London

Stansfield GRV (2007) Iraq: people, history, politics. Cambridge, Polity

Tripp C (2007) A history of Iraq. Cambridge University Press, Cambridge

Tucker S, Roberts PM (2010) The encyclopedia of Middle East wars: the United States in the Persian Gulf, Afghanistan, and Iraq conflicts, Santa Barbara. Calif, ABC-CLIO

United Nations Educational, Scientific and Cultural Organisation (2014) Erbil Citadel [Online]. Online: United Nations Educational, Scientific and Cultural Organization. http://whc.unesco.org/en/list/1437. Accessed April 2015

Watson GB, Bentley I (2007) Identity by design. Amsterdam; London, Butterworth-H

Yaqoobi D, Michelmore D, Tawfiq R (2012) Highlights of Erbil Citadel. High Commission for Erbil Citadel Revitalization, Erbil

Chapter 10
Post-war Restoration of Traditional Houses in Gaza

Suheir M. S. Ammar and Nashwa Y. Alramlawi

Abstract In the past decade, the city of Gaza has suffered from three consecutive wars in 2008–2009, 2012 and 2014. As a result, severe physical destructions impacted all aspects of urban life. The impact of destructions on historical buildings is undoubtedly of the most significant, because it represents a destruction of the identity and cultural history of Palestinian people, which is part of the world heritage that belongs to all. This chapter discusses a collaborative project between a local institution, the Architectural Center for Heritage (IWAN) at Islamic University of Gaza (IUG), and an international organization, the International Committee of the Red Cross (ICRC), for the post-war restoration of traditional houses after the 2008–2009 war that was implemented in collaboration with the residents of those houses. It provides a local perspective on a project-targeted 37 traditional houses in the Old City of Gaza that was partially affected by the war of 2008–2009. The study presents the stages of the work including documentation, awareness raising workshops, training, the cleaning phase and the restoration works. In addition, the study presents the challenges that faced the project, and how they were overcome. Financing was one of the challenges as it was limited. There was a need to define priorities of the most important works to be done. Another one was the lack of skilled workers and engineers which required training for them. Additionally, implementing the conservation works carried out while residents inside the house. The results show that it is possible to get a great achievement with limited resources. The project finance was for temporary job creation, and the partnership invested it to conserve the affected traditional houses. As a result of community participation, residents got awareness about the cultural value of their houses, and how to maintain them. This knowledge about the mechanism of maintenance contributed to its sustainability for future.

Keywords Restoration · Post-war recovery · Traditional houses · Gaza

S. M. S. Ammar (✉) · N. Y. Alramlawi
IUG, Gaza, Palestine
e-mail: sammar@iugaza.edu.ps

N. Y. Alramlawi
e-mail: nalramlawy@iugaza.edu.ps

© Springer Nature Switzerland AG 2020
F. F. Arefian and S. H. I. Moeini (eds.), *Urban Heritage Along the Silk Roads*,
The Urban Book Series, https://doi.org/10.1007/978-3-030-22762-3_10

10.1 Introduction

Traditional buildings are parts of the cultural heritage of nations, and conserving them represent conserving the identity of these nations. Félix et al. (2013) stated that a house is not only a space to live in, but it gives a sense of dignity, belonging and cultural identity. They demonstrated that the solutions that respect cultural and local conditions are more successful in post-disaster housing.

Gaza is considered one of the oldest cities in the world. Since 3000 years ago, there were many civilizations in Gaza such as Canaan, Greek, Roman and Islamic civilization. Gaza still has archaeological sites and traditional buildings from these civilizations including houses, markets, mosques, public baths (Almubaed 1995).

National plan for traditional architecture (nd) in Palestine defines the traditional architecture as 'the range of constructions derived from the rooting of a community within its territory, revealing in its diversity and evolution a process of ecological adaptation both to natural resources and factors and to the historical processes and socio-economic models which have developed in each location. They represent a key reference point among the cultural hallmarks of the community generating them, as the result of shared knowledge and experience, transmitted and enriched from one generation to the next'.

The traditional houses targeted by the project were houses dating back to the Mamluk and Ottoman periods. They are located in the heart of the Old City of Gaza, which includes four main neighbourhoods, namely, Addaraj, Azzaitoon, Attofah and Ashajaiah (Fig. 10.1).

Addaraj neighbourhood is the largest and the oldest neighbourhood in the old town of Gaza, and it is located on the city's hill. Azzaitoon neighbourhood is also considered as one of Gaza's old neighbourhoods. It is to the south of Omar al-Mukhtar Street and Addaraj neighbourhood is located to the north of Omar Al Mukhtar Street. Attofah neighbourhood lies north of the old hill, and it is called attofah, which means apple, because it had many apple trees in it. Ashajaiah neighbourhood is named after martyr leader Shujauddin Othman al-Kurdi, and it is located out of the boundaries of the old hill to the east of the three other neighbourhoods (Almubaed 1987).

Traditional houses constitute the largest number of traditional buildings that still exist within Gaza's Old City. The war machine destroyed many buildings in these four neighbourhoods including the historic Municipal Court in Addaraj.

This chapter presents an experience of collaboration between IWAN Centre and the International Committee of the Red Cross (ICRC) to restore the affected traditional houses in Gaza after the war of 2008–2009. It highlights the impact of war on cultural heritage, gives a general description of the traditional houses in the old town, explains the collaboration between IWAN Centre and ICRC including the stages of work in the project and underlining the role of community participation. It ends with recommendations. The project included 20 houses in Azzaitoon, 8 houses in Addaraj and 9 houses in Ashajaiah neighbourhood. The fourth neighbourhood was not included.

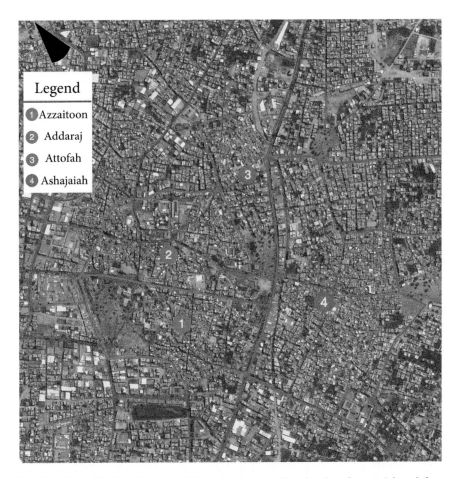

Fig. 10.1 Gaza Old City layout including the four neighbourhoods. *Source* Adapted from Almoghany (2007), p. 50

10.2 Impacts of War on Cultural Heritage

The effect of wars on cultural heritage could be from directly destroying them completely to partially. Yet, this is not the only effect. The indirect effect through targeting adjacent buildings may cause cracks or landslides as a result of explosions which cause strong vibrations. The degree of seriousness of cracks varies depending on the explosion strength and the building stability. Non-visible damages are serious and require care and accuracy when diagnosing damages.

On 27 December 2008, Israel launched 3 weeks of intense military bombardment on the Gaza Strip, killing nearly 1,500 people. The combination of war and siege totally devastated the lives and infrastructure of Gaza. During this war, 3,550 houses were completely demolished, 50,000 houses were partly damaged, out of

Fig. 10.2 The compact fabric of houses in the Old City. *Source* Authors

which 13,000 ones were unsuitable for living (Ministry of Public Works and Housing 2009).

During the war of 2008–2009, many places, including houses and governmental buildings, in the Old City in Gaza were targeted and destructed which had a great impact on the nearby historical buildings. The compact fabric and adjacent houses without setbacks have protected many of these buildings (Fig. 10.2).

Additionally, there were some damages in some of these buildings mainly, due to the severe vibration resulting from powerful bombs which have high destructive techniques. The impact of vibration and damages to the historical buildings in this case, can be concluded as follows.

- The appearance of large clear cracks in some buildings which were not visible before the war created great fear among the residents of these houses. There was an urgent need to intervene and treat them.
- Increase in the size of some old cracks in some buildings because of the soil movement that resulted from the strong nearby strikes.
- The emergence of some minor cracks that does not embody a risk, but need a simple treatment to reassure the inhabitants of the houses.
- The appearance of some cracks in the roofs of the upper floors in some houses, causing the leakage of rainwater into the inner part of the house (Fig. 10.3).
- There were some unseen damages that need to be examined to avoid any danger in the future and to preserve the sustainability of the building.

Fig. 10.3 The cracks that appeared after war. *Source* Authors

Importantly, during the war, the traditional Municipal Court building in Addaraj neighbourhood in the heart of the Old City suffered extensive damages. It could be observed that almost half of it was destroyed. There were also complete destruction and severe damages to residential buildings, police stations and governmental buildings within the boundaries of the Old City.

The appearance of such new damages, the increase of old damages and the presence of latent damage in these traditional buildings are considered as great threats to the existence of these traditional buildings. At the time, there were serious threats from some owners to demolish their houses to avoid any risks. Therefore, traditional buildings need maintenance regularly, especially, in areas where war is expected to avoid increasing damage. Wang and Li (2006) and Yau (2009) argued that proper maintenance is essential to protect people's well-being and to prolong the housing structures' life which helps to delay new structure redevelopment needs. Cheong (2010) argued that neglecting the maintenance of a building can affect its value and age; regular maintenance will keep it attractive in the market. The three authors have spoken about buildings in general; however, the traditional buildings need more maintenance than other buildings for their cultural and historical values.

Accordingly, there was a need to identify the damaged buildings, assess the damages, the seriousness of damages and identify the causes of damage. These points were defined as a priority to preserve the traditional buildings in the Old City of Gaza.

10.3 Characteristics of Traditional Houses in the Old Town of Gaza

Most traditional houses in Gaza have similarities in architectural elements and characteristics. Providing privacy for the family is the most important goal besides a comfortable living environment. The most important feature in most of these traditional houses is the inner courtyard, which represents the heart of the house and the main element which is surrounded by other spaces (Fig. 10.4). Most of the daily activities occurr inside this courtyard. It is the source of natural light and ventilation for the surrounding spaces.

An external backyard is sometimes available. Most openings are oriented towards the courtyard and are large enough to provide natural lighting and ventilation. Nevertheless, some small windows, higher than the level of passers-by eyes, are oriented towards the street and covered with wooden oriel 'mashrabiat'.

The indirect narrow entrance vestibule with low ceilings protect residents from being seen from outside when the entrance door is opened (Fig. 10.5).

Fig. 10.4 The inner courtyard in a traditional house

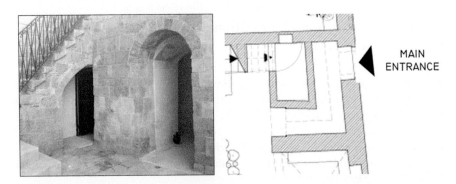

Fig. 10.5 The indirect narrow entrance vestibule

Furthermore, the Iwan which is a enclosed space opening towards the courtyard represents a distinct element that serves as a living room in most of these houses (Fig. 10.6). Another important element called Almazirh, near the entrance, is used in many of these houses to conserve water that is transferred to the house from the closest water wheel. This water is used by residents for drinking and cooking (Fig. 10.7).

Fig. 10.6 An Iwan

Traditional houses have stone decorations on their interior facades, especially on the top of the openings, doors and windows (Fig. 10.8). These decorations are usually of arabesque or geometrical configurations. They are precisely drawn patterns and reliefs made of limestones or marble.

The construction system of traditional houses is usually load bearing walls made of local materials such as natural stone, clay and pottery. They are mostly consisting of one or two floors. There are recessed niches called Youk, used for keeping residents' possessions. The roofing system consists of cross-vaults from inside topped by domes (Fig. 10.9). However, flat roof is sometimes used with appropriate technology.

Traditional houses vary in their floor areas and ornaments depending on the economic situation or social status of the owner. Some are small single-storey houses while others have two storeys. Each house has a large inner courtyard and several Iwans for sitting. Some of these houses have the first floor added in a later period. Some have a first floor of modern concrete blocks that is completely different.

In spite of these design differences in the formation of traditional houses, the philosophy of design and common elements depend on the same principles that characterised cities of the Middle East at that time.

Fig. 10.7 Almazirh

Fig. 10.8 An example of decorations

Fig. 10.9 The cross-vaults of ceiling

The appearance of some of these buildings was changed as a result of residents' interventions. In his study about Gaza, Muhaisen (2009) clarifies this, and states that there are many wrong techniques in maintaining these buildings including using modern materials like cement plastering for walls and ceilings from inside and outside. In other cases, they used tiles for walls in kitchens and bathrooms. The worst was adding other rooms beside or over the house from modern construction materials. This intervention prevents the needed ventilation for sandstone walls. Field studies showed that the main reason for not maintaining most of these houses is lack of finance. Also, Muhaisen and Alheta (2008) state that there is a lack of qualified workforces in the field of conservation in Gaza.

Praising the traditional structural system, Jigyasu (2013) argues that the performance of traditional buildings using local building techniques is much better than many badly built modern structures. Talking about post-earthquake damages, he observes that although there might be cracks in most cases, but traditional buildings do not collapse minimising loss of residents' lives. However, adding reinforced concrete floors about traditional buildings, he observes, is a major reason for extensive damages during disasters. These additions are the result of a lack of adequate knowledge among residents. Post-disaster reconstruction usually faces the problem of unavailability of skilled workforce to conserve traditional buildings. He asks for preventing the replacement of traditional building after a disaster with modern ones, and maintaining the local sense of identity on the accumulated experience from the past. Earthquake causes vibration akin to bombing and causes similar damages to traditional buildings.

10.4 The Project

Immediately after the war, relief institutions and organisations began to work actively in the Gaza Strip in order to provide shelters for the families whose houses were demolished. The priority of the humanitarian response to this war was oriented towards financing health, children, women, nutrition and education sectors. The cultural sector including cultural heritage was not a priority nor significant for donors. It was to some extent a kind of luxury in the light of human and environmental damage.

After a brief period following the war, temporary work projects started to reduce poverty and high levels of unemployment in the area. They aimed to create jobs for some unemployed poor labourers, whose numbers increased after the war. Many institutions, such as the International Committee of the Red Cross (ICRC), provided employment opportunities ranging from 3 to 6 months per person to provide as many working opportunities as possible such as removing wrong interventions, filling the cracks, traditional plastering and pointing stone blocks called Kohla, and insulation and finishing for the roofs.

As a local institution, IWAN Centre thought of orienting the temporary work programmes towards conserving the cultural heritage, and at the same time preserving the living environment of the inhabitants of traditional houses. The project was born after signing an agreement between the IWAN Centre and the International Committee of the Red Cross. The idea was to provide materials, tools and a working diary for about 230 workers, craftsmen and engineers dividing the finance by 40% for materials and tools and 60% for workers, craftsmen and engineers.

The ICRC provided workers, craftsmen and engineers and building materials as needed. Because of the siege on Gaza, the ICRC had also to coordinate to provide necessary materials and tools through the crossing's borders. Concurrently, IWAN Centre selected the houses to be maintained based on the structural status of the house, the economic situation of the residents and the nature and degree of the damage.

10.4.1 The Project's Work Stages

10.4.1.1 Needs Assessment

The project team made six visits to the houses selected for conservation works to decide about the required interventions and the works to be carried out for each house (Fig. 10.10). Then, they determined the amount of materials and tools, the number of needed workers, technicians and engineers for each house.

Fig. 10.10 A visit from IWAN and ICRC

10.4.1.2 Training Phase

The project offered temporary employment opportunities for the unemployed workers and engineers, who did not have prior experience in the conservation of the traditional buildings. Therefore, a training workshop was organised for six days for the engineers to provide them the necessary skills to supervise the conservation projects, and how to manage the workers and materials inside the site without causing any damage to houses (Fig. 10.11). This was necessary to get a good quality of work in the project as required.

Fig. 10.11 A training workshop

10.4.1.3 The Restoration and Direct Intervention Phase

Technical works on the restoration of the traditional houses began including the following items: identifying the locations of visible cracks, implementing a good structural reinforcement to them and then filling the cracks and injecting them to stop the damage resulting from them (Figs. 10.12 and 10.13).

10.4.1.4 Removing Wrong Interventions Phase

Wrong interventions, which were made in the past by the residents of houses who do not have sufficient awareness of the nature of the interventions that are appropriate to the standards of conservation and restoration of traditional buildings were removed (Fig. 10.14). Most of those interventions were cement layers to fill cracks

Fig. 10.12 A visible crack

Fig. 10.13 Filling the cracks

Fig. 14 Removing the cement layers

and conceal damp areas. In some buildings, ceramic tiles were used to cover walls in kitchens and bathrooms using concrete mortar. In addition, external random electrical installations that affect the aesthetic elements of historical built environments were found. There was a need for supporting the building parts during this stage.

10.4.1.5 The Finishing Works Phase

During the removal of all the wrong interventions, there was a need for quick treatments for some problems that occurred. After the removal works, the finishing works began, and they included the work of traditional plastering and filling the joints between stoneskohla, the supervision team defined the places of plastering and others for kohla (Figs. 10.15 and 10.16).

Fig. 10.15 Filling the joints between stones

Fig. 16 Plastering

10.4.1.6 Insulation and Exterior Finishing Works' Phase

The next stage was the works of insulation and exterior finishing for the roofs to address the problems of water leakage in some houses and to prevent moisture and any leaks that may occur in the future.

The implementation of these works has contributed to improve the living environment of the inhabitants of the traditional houses, and stop the damages and danger that could have increased as a result of delaying the intervention (Figs. 10.17a, b and 10.18a, b).

10.5 Residents' Participation

The project is distinguished by effective community participation during the implementation of the whole works, which some have considered important. Leung (2005) and Sanoff (1990) stated that residents' participation in any works in their houses creates a feeling of satisfaction. Sanoff added that this feeling is a result of fulfilling requirements and is related to the feeling of having influenced the decisions.

This participation from a few of the family members was observed in the following items:

- Residents participate in determining places of intervention in their houses and actually, they had fluenced the decisions. In addition, they took the responsibility of executing some works that were not included in project budgets such as electrical installation and sanitary works. This in turn contributed in adding to the future value of their houses.

Fig. 10.17 a, b Before restoration

- Facilitating the work of the labourer teams inside their houses and offering a good working environment for workers and engineers who contributed to the success of the work and speeded the implementation of works according to design.

Residents acquired a good experience about appropriate intervention mechanisms for traditional buildings as a result of discussions between them and the supervision team. This contributes to the sustainability of the project as the residents get experience in conserving their houses correctly, in case they need maintenance work in the future. In addition, a good relationship was shaped between them and IWAN Centre as a specialised centre in architectural heritage to ask for any technical support in the future.

Fig. 10.18 a, b After restoration

10.6 Recommendations

In the light of the study, there are some recommendations:

1. Residents' knowledge about their houses' value and how to conserve them is important and not difficult achieve.
2. Residents' participation in the process of conservation from the outset is important and increases their satisfaction about the work.
3. Creative thinking can help exploit financing opportunities.
4. Responsible authorities and associations should be liable for the poor families who do not know how to fulfill their needs.
5. While implementing unemployment programmes, it is possible to provide workers with special skills that can grant them future work opportunities.

10.7 Conclusion

The war affected the traditional houses which represent the identity and culture of the Gaza Strip. The post-war financing cares more about health, children, women and education sectors. Although post-war reconstruction delayed the value of the traditional buildings and their limited number make their conservation a priority. This chapter described the impact of war on traditional houses in Gaza, including the appearance of large visible cracks which worried inhabitants, the increase in the size of some old cracks, as well as new cracks, with some in roofs causing rainwater leaks. The collaboration between local and international organisations focused on the phase of implementing the project beginning with need assessment, training, reinforcement, removing the wrong interventions, traditional plastering and filling the joints, insulation and exterior finishing for the roofs. It also showed that residents' participation in conserving traditional buildings was encouraging in this project. They participated in defining the places of intervention in their houses and support financially some works by taking the responsibility of executing some works that were not included in the project budget. Additionally, residents of these houses acquired a good experience about the mechanism of appropriate intervention on traditional buildings which can help them for maintenance in the future.

References

Almubaed S (1995) The Islamic historical buildings in Gaza and its strip. The Egyptian General Authority for Books, Cairo, Egypt

Almubaed S (1987) Gaza and its strip. A study in the immortality of the place and the civilization of the population. General Authority of the Book, Cairo

Almoghany N (2007) Architectural heritage in Gaza City. Riwaq-Centre for Architectural Conservation, Ramallah, p 50

Chu Ka Cheong W (2010) The impact of owners' participation in quality private residential housing management (Master. Sc.). The University of Hong Kong. http://hdl.handle.net/10722/128611. Accessed 10 Mar 2017

Félix D, Feio A, Branco JM, Machado JS (2013) The role of spontaneous construction for post-disaster housing. In: Structures and architecture: concepts, applications and challenges, pp 937–944

Jigyasu R (2013) Using traditional knowledge systems for post-disaster reconstruction-issues and challenges following Gujarat and Kashmir earthquakes. Creative Space (CS) 1(1):1–17

Leung CC (2005) Resident participation: a community-building strategy in low-income neighbourhoods. Joint Center Housing Stud Harvard University, Cambridge, p 38

Ministry of Public Works and Housing (2009) Losses in the public and private installations in the Gaza Strip due to the Israeli aggression

Muhaisen A (2009) The reality of historical residential buildings in Gaza city and the ways to preserve them. The Islamic University J (Series of Natural Studies and Engineering) 17(1):109–132

Muhaisen A, Alheta D (2008) A study of the reality of the technical rehabilitation of the workers in the field of architectural preservation in the Gaza Strip. Paper presented at the Architectural Heritage Conference. The reality and challenges of conservation, Gaza, Palestine

National plan for traditional architecture (nd) National plan for traditional architecture

Sanoff H (1990) Participatory design: theory and techniques. http://books.google.com/books?id=vedEAAAAYAAJ. Accessed 20 Mar 2017

Wang D, Li S-M (2006) Socio-economic differentials and stated housing preferences in Guangzhou, China. Habitat Int 30(2):305–326. https://doi.org/10.1016/j.habitatint.2004.02.009

Yau Y (2009) Weightings of decision-making criteria for the maintenance of multi-storey residential buildings in Hong Kong. Paper presented at the European Network of Housing Research Conference, Prague, Czech Republic

Part IV
Zooming In: Urban Heritage in Iran

Chapter 11
Traditional House Types Revived and Transformed: A Case Study in Sabzevar, Iran

Karin Raith and Hassan Estaji

Abstract This chapter is not primarily concerned with the preservation of historical buildings—for which it is often too late—but rather with the question of how continuity between the cultural heritage and contemporary architecture can be established. In Sabzevar, Iran, many culturally valuable residential buildings have been demolished in recent decades due to the rapid economic and physical growth of the city and profound social changes. As real estate prices have risen, the density in inner-city areas has also increased, negatively impacting the historic urban fabric of low-rise courtyard houses. The traditional extended families have been gradually replaced by small households who prefer small apartments. Thus, it seems that the evolution of autochthonous house types has ended and that in the future only global standard housing will be constructed. An analysis of fourteen listed residential buildings revealed their careful adaptation to the desert climate. The oldest buildings, least influenced by European architecture, provided the best thermal comfort. Nowadays, however, electric air-conditioning is preferred to traditional temperature management. Although many houses could be adapted to new ways of living and working, the prospects of financial profit by rebuilding a property often outweigh the appreciation of a building's cultural significance. While architectural heritage is neglected in favour of progress, the loss of local identity is being mourned. The paper highlights the surprising potential of two traditional house types to be transformed, typologically developed and applied to new urban developments, and it presents arguments for their revival:

- Time-tested environmentally adapted structures help saving energy.
- A flexible layout and neutral spaces provide the best options for an adaptation to new lifestyles.
- A reinterpretation of traditional typologies by use of advanced construction methods and contemporary design vocabulary enhances the local character.

K. Raith (✉)
University of Applied Arts Vienna, Vienna, Austria
e-mail: karin.raith@uni-ak.ac.at

H. Estaji
Hakim Sabzevari University, Sabzevar, Iran
e-mail: h.estaji@gmail.com

Keywords Cultural heritage · Environmental adaptation · Flexibility · Reinterpretation · Traditional house typologies · Sabzevar

11.1 Introduction

The residential building stock of Sabzevar, a city on the northern rim of the Central Iranian Plateau, was once composed of large courtyard houses. The impressive buildings accommodated large families and were perfectly adapted to the harsh desert climate. But in the last decades, most of these culturally valuable urban houses have been torn down. This was the result of three major developments:

Firstly, the city has been rapidly expanding due to economic growth. Real estate prices in inner-city areas have skyrocketed, thus making it profitable to demolish the low-rise courtyard houses and to rebuild the plots with multistorey apartment or office buildings.

Secondly, lifestyles have diversified. The traditional sedentary way of life in large families has been gradually replaced by geographic and social mobility and small households. Only wealthy families can afford the spacious old-style residences, while the majority of Sabzevar's inhabitants prefer to live in smaller modern apartments.

Thirdly, a change in climate control is posing a threat to traditional architecture. In the last decades, fossil energy has been extremely cheap and only recently has sustainability started to become an issue in architecture. Therefore active ('artificial') HVAC systems are now generally preferred to passive ('natural') cooling and heating. The habit of moving to different rooms within the courtyard houses depending on the season and the time of the day has been abandoned; people nowadays live in compact, artificially climatised flats modelled after Western buildings. This means that the space-consuming provision of thermal comfort by specialised summer and winter zones within the house has been widely replaced by an energy-consuming system.

It seems that the evolution of autochthonous house types has ended and that in the future only apartment buildings will be constructed that conform to global standard schemes. This development is controversial in many respects: modern Western-style residential architecture with balconies and large windows enjoys high prestige but it does not adequately meet the needs for privacy deeply rooted in Iranian culture. Although the value of the few remaining examples of historical building culture is officially recognised, governmental policies do not support their preservation and maintenance either on a legal or a financial basis. While cultural heritage is carelessly dismissed in favour of progress, the loss of local identity is mourned. These paradoxes could be—at least partly—resolved by identifying the potentials and deficits of traditional buildings and by reproducing their qualities in contemporary architecture.

11.2 Traditional Courtyard Houses in Sabzevar

A survey and examination of fourteen residential buildings in Sabzevar (Estaji 2017) provided the basis for the analysis as to whether their elements and typological conception would qualify them for a relaunch in the historic city and urban expansion zones. The researched houses (Fig. 11.1) are all registered on the National Iranian Heritage list and represent the last examples of the autochthonous architectural style (Isfahani) before the emergence of Modernism in Iran. Since Sabzevar is a small historical city, far from the capital Tehran, the transition from traditional to modern architecture did not take place before the first Pahlavi period (1925–41).

11.2.1 The Urban Context

The water supply system was one of the crucial factors that influenced the morphology of the Iranian desert cities. As Sabzevar grew and farmlands were transformed into urban lots for residential buildings, the maintenance trails along the

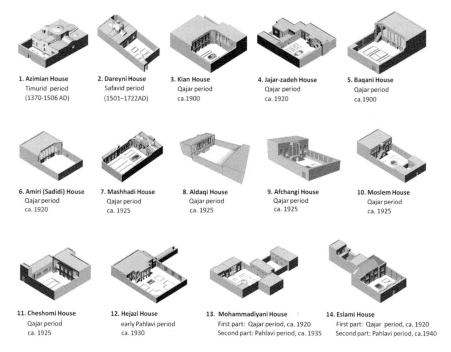

Fig. 11.1 Traditional houses of Sabzevar, sorted by building age (from top left to bottom right) (Estaji 2017)

water channels were turned into streets. The channels disappeared, but the urban pattern of old Sabzevar remained an accurate representation of the irrigation network; even the hierarchy of streets followed the hierarchy of the water distribution system (Estaji and Raith 2016). Since the terrain slopes from north to south, the direction of the main channels was the same, with secondary channels laid out in perpendicular direction. Research into the energy performance of the fourteen listed buildings showed that the best direction for the main rooms to minimise solar gain in summer and maximise it in winter is south (Estaji 2017)—so there is a perfect convergence between the orientations of these two systems.

This historically developed structure was profoundly changed, when around 1975 the municipalities commissioned a master plan for Sabzevar with the goal of widening the streets for individual motor car traffic. New streets cut into the urban fabric, some of the old buildings were demolished and the layout of existing neighbourhoods changed. The second transformation phase began around 2000 when the common types of courtyard houses were replaced by multistorey apartment blocks (Estaji and Raith 2016).

11.2.2 The Social and Economic Context

Until the Qajar period a household consisted of the extended family that comprised several generations living together in a large courtyard house with shared kitchen, storage and service rooms. When the family grew a new wing could be added to the building and servants could be accommodated thanks to spatial reserves offered by the courtyard. Large houses could also be divided into smaller units by locking doors or blocking openings.

Before the Pahlavi period most of the families in Sabzevar and in other central Iranian cities subsisted on agriculture (Kheirabadi 2000; Mahdavi 2009). The basements and side wings of the houses were required for storing agricultural products and equipment. By the end of winter, seeds and tools were moved to the fields that were usually located near the city or in the ancestral villages of the residents. Surplus agricultural products were sold at the market. The emptied rooms were then transformed to accommodate other usages. The compact mode of life in the cold season could be converted into a distributed living in summer (Kheirabadi 2000; Mahdavi 2009). The current changes in lifestyle and the economic shift from agriculture to the secondary and tertiary sectors have led to different spatial requirements.

11.2.3 Flexibility of Traditional Houses

The flexibility of the courtyard houses resulted not only from their potential to be expanded and divided, but also from two further factors:

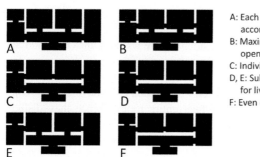

Fig. 11.2 Some possible scenarios for the first floor of the Aldaqi house (Estaji 2017)

Multifunctionality of spaces: Rooms were sparsely furnished and thus equally suitable for living, dining, sleeping, resting and socializing (Drouville 1819; Mostowfi 1942/1997). Living spaces were at the same time circulation spaces.

Maximum connectivity between rooms: Due to doors between all adjoining spaces, the configuration of rooms could easily be changed, as shown in Fig. 11.2 (Estaji 2017).

Some of the investigated houses have experienced a lot of changes. A wing of the Aldaqi house (Fig. 11.1, #8), for example, was converted from storage space into commercial premises. The Kian house (Fig. 11.1, #3) was transformed into a bank branch about one hundred years ago. The functions of the Mashhadi and Eslami houses (Fig. 11.1, #7 and #14) changed several times (Estaji 2017). Multifunctionality is an important factor in extending the lifespan of buildings and saving them from demolition. As long as cultural concepts and lifestyles developed at a slow pace, traditional houses were able to respond to these changes by modifying the relationship between spaces. A sudden increase in population, rapid economic growth, and change in family structures have challenged this capacity.

11.2.4 Climatic Adaptation

Although the best orientation of the main rooms regarding energy efficiency and living comfort is to the south, many traditional houses in Sabzevar face east, due to constraints resulting from the proportion of the plots and the way they are embedded in the urban fabric. The disadvantages in terms of temperature management were compensated with particularly deep verandas built in front of the east facades. They protected the main rooms facing the courtyard from solar radiation at a low angle in the morning and at the same time provided a comfortable living space for summer evenings, in spring and autumn (Estaji 2017). The other exterior walls were either attached to neighbouring houses or windowless in order to minimise solar gain and ensure privacy. All houses were thus inward looking and surrounded by the dense urban fabric.

Originally, rooms had no heating. The residents wore warm clothes in winter and took advantage from a microclimate comfort zone provided by a 'Kursi', a low wooden table on a charcoal brazier, surrounded by mattresses and cushions and covered with a large thick quilt under which all huddled (Mahdavi 2009). At the end of the Qajar period people started to heat entire rooms. Since they disliked entering cold semi-open and open spaces when they moved between heated rooms, internal corridors were introduced around the Pahlavi period. Verandas thus lost their function as circulation spaces, and were downsized and reduced to mere decorative elements (Estaji 2017) (Fig. 11.3).

Due to the low price and abundant supply of oil since the Pahlavi period and the use of active heating and cooling systems, the complex passive strategies of traditional architecture have been abandoned. Roofed semi-open spaces used for

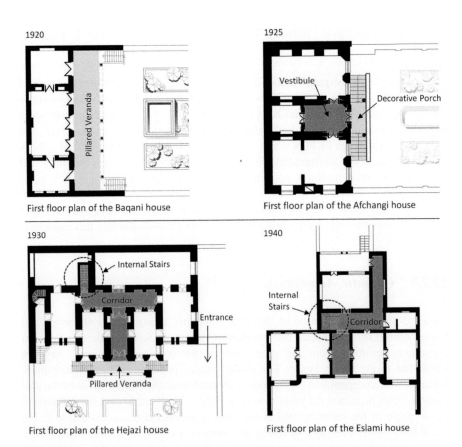

Fig. 11.3 Evolution of circulation spaces and transition zones from the courtyard to the building (Estaji 2017)

| Mashhadi house | Afchangi house | Eslami house |
| End of Qajar period | End of Qajar period | First Pahlavi period |

Fig. 11.4 Changes in the size, proportions and positions of windows under the influence of European architecture compromised the thermal performance of the buildings. *Source of photos* Estaji (2017)

shading gradually disappeared. Thus, facades were exposed to direct sunlight—a particularly adverse development, because under the concurrent stylistic influence of European architecture, also the size and proportions of windows changed (Fig. 11.4). Small windows recessed deep into the reveals were replaced by wide windows flush with the facade. The posts between the traditional window-doors of the 'seh-dari' and 'panj-dari' (three- and five-door-rooms) that had served as vertical shading devices—a sort of brise-soleil—had effectively blocked solar radiation from the side and at a low angle in the mornings and evenings. The large amount of energy absorbed by glazed areas changed the thermal performance of the buildings for the worse. Therefore, the houses from the late Qajar and the Pahlavi periods are warmer in summer and need active cooling systems for acceptable indoor temperature conditions.

11.3 Unpredictable Future Developments Demand Versatile Buildings and Mixed-Use Urban Quarters

With many ongoing socioeconomic transformations, residential architecture has to transform as well. Rather than merely responding to present-day developments, it is necessary to design flexible building types that are capable of accommodating future changes without major structural alterations.

In the past, the internal and external social relations were clearly defined: the houses were designed for extended families and their guests, the private ('andaruni'), semi-private ('biruni'), and public spheres being strictly separated (Nafisi 2002). Traditional courtyard houses were able to accommodate various changes in terms of the number and constellation of residents as long as these happened among families, but they could not easily be subdivided and shared among unrelated persons without violating privacy. This means that as soon as the underlying social construct became obsolete the flexibility of the physical construct also hit its limits.

Today, relatives still occasionally live together, with the variety of communities sharing an apartment or house has becoming more multifaceted. In the future, it will include single households, supervised residential accommodation for elderly persons, student shared households and other residential relationships. Not only this multitude of living arrangements has to be accommodated, but also the segregation of residential and business architecture based on the modern functionalist doctrine has to be overcome in favour of more versatile and thus sustainable mixed-use urban quarters. In fact, many of the traditional houses once included utility rooms, commercial spaces, storage rooms and workspaces of various kinds. This diversity of use must again be made possible by an appropriate spatial offer.

Separating the private sphere of family rooms and private open spaces from the semi-private guest area and the provision of different spaces for hot and cold seasons meant that there was once a clear definition—and thus specialisation—of zones within the house in terms of privacy and thermal performance. On the other hand, the rooms within these specialised areas were designed as versatile multi-purpose spaces (Mahdavi 2009). Hence, the multifunctionality of the rooms compensated for the high space requirements for separate winter and summer zones as well as family and guest areas. When Western architecture gained influence in Iran, the traditional courtyard house was confronted with an almost diametrically opposed concept of residential architecture. In European middle class housing, there is neither a strict segregation of private and guest zones, nor a comparable differentiation of thermal comfort zones, but rather a pronounced functional specialisation of rooms with specific furniture and different room dimensions (especially in modernist architecture). With the adoption of Western furnishing practices, the flexibility of Iranian living culture was destroyed. The conflict resulting from the fusion of Western concepts with Iranian domestic culture can only be solved when the specialisation of spaces is reduced again to allow for more diverse uses.

These social and economic changes and the associated architectural challenges cannot just be observed in Sabzevar—the city is rather a prime example of developments taking place throughout the entire region of central Iran. Due to increasing land prices in inner-city areas there is a demand for building types that allow for higher densities. Spatial compositions are required that can be stacked to form multistorey residential or mixed-use houses. Nevertheless, essential qualities of the premodern courtyard houses should be preserved. Reviving particular elements and typological characteristics of traditional houses will also help to maintain the city's identity.

11.3.1 Qualities of Traditional Houses

There are several elements of traditional courtyard houses that provide essential qualities. They should be re-introduced into contemporary architecture in a modified, modernised form:

Courtyards offer clear advantages in terms of building physics and climate control. The inward oriented conception of courtyard houses allows for high-density low-rise development, reducing exposed walls and thus avoiding solar gain in summer and heat losses in winter. Solar radiation on the facades facing the courtyards can be controlled by verandas. Courtyards generate favourable microclimates, they protect against hot and dusty winds, preserve air humidity and—especially when planted—provide shade and evaporative cooling. As private open spaces, they once represented intimate extensions of the living rooms, but even when transformed to semi-private areas in the case of multistorey buildings they could provide quiet outdoor places secluded from traffic noise and urban bustle.

Verandas ('eivans'), placed in front of the facades, play a vital role for passive climatisation. They serve to control solar radiation on the outer walls, protect them from the hot summer sun, but invite the desired winter sun into the rooms. Their function as thresholds between interior and exterior could be further improved by movable louvers, folding shutters, 'mashrabiya', sliding glass elements and other sorts of screens that provide privacy as well as protection from sun, wind, heat and cold and establish a semi-open zone whose permeability can be adapted to changing requirements. Activated in this way, the veranda could also form a buffer zone into which the adjoining rooms could temporarily expand. Thus, spatial flexibility would be further enhanced.

Generic rooms can accommodate various uses allowing for organisational flexibility. The more the composition of spaces is customised to suit particular needs and the room character is tailored to specific requirements, the more difficult it is to adapt the building to changing necessities. Symmetries that are common in traditional houses add to prestigious aesthetics but make it more difficult to reconfigure the rooms. A modular composition of rather neutral, unspecific spaces facilitates conversion. Rooms with direct access from circulation spaces offer more options than spaces that have to be crossed to enter another room. The most versatile buildings are those that do not only serve residential purposes but are also able to provide workspace.

'Predetermined breaking points', i.e. potential openings in the dividing walls enable the residents to combine spaces to larger room entities or subdivide them to smaller units. The traditional segmentation of the wall into a series of structural pillars and non-load-bearing partitions or niches requires only minimal interventions to provide doors or other openings.

11.3.2 Deficits of Traditional Houses

Some characteristics of traditional courtyard houses are incompatible with contemporary economic and social conditions. The climate-responsive performance of traditional houses was partly based on the idea of moving to that part of the building which offered the most favourable thermal conditions—a principle that was reflected in the term 'four-season house'. This meant that large parts were temporarily unused or underused. These ample spatial reserves cannot be provided under the present economic and social conditions. Instead of reproducing the specialised summer and winter *rooms* themselves, today the *effects* created by them have to be reproduced. Subterranean spaces, for example, provided a comfortable cool indoor climate in summer due to the thermal inertia of the surrounding earth mass which has a stable temperature near the average annual air temperature. The same effect can be produced by geothermal cooling with ground source heat pumps. Technical progress allows to use passive heating and cooling systems without the comprehensive spatial provisions that were necessary in former times and without the use of fossil energy.

11.3.3 House Types to Be Revived and Transformed

The analysis of the listed buildings allowed to distinguish different house types which are not equally suitable to meet contemporary and unpredictable future needs. Two of them could be further developed and modernised.

11.3.3.1 The Chahar-Soffeh Type

The Azimian house (Fig. 11.5), a chahar-soffeh (chahar-soffa) type from the Timurid period, is inward oriented and enclosed by nearly windowless walls. Originally it was integrated into a dense urban fabric. Decades ago the stable and storage rooms on the south side of the courtyard were demolished for road construction, but the main wing is still inhabited and kept in good condition. The current owners carried out small conversions without destroying the integrity of the house: they added a bathroom and toilet and reduced the kitchen size instead. This demonstrates that even the oldest houses—carefully maintained—would meet today's comfort standards. However, living in such a house requires cultural interest, investment in the conservation of the building and a readiness to adapt one's lifestyle to its character, because the rigid symmetrical geometry makes it a rather inflexible building type.

The four-eivans plan with its cross-axially arranged vaulted spaces that open either to a central domed space or courtyard (Fig. 11.6) had already been developed by the Parthians and Sassanians for their palaces and temples (Keall 1974, p. 124).

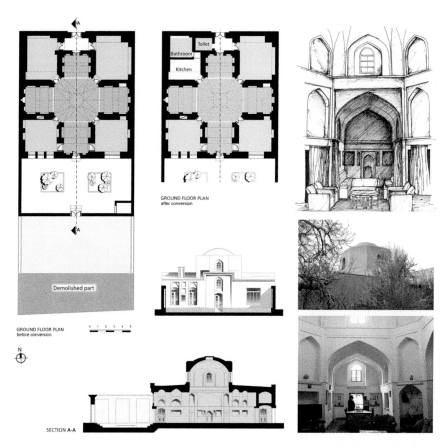

Fig. 11.5 Azimian house (Estaji2017, based on Towhidi-Manesh 2002; hand drawing by Minoo Ghasemi)

Since the twelfth century the scheme has been widely applied to Islamic structures of various scales such as mosques, madrasas and caravanserais. Highly formalised for stately and religious architecture, the building type is less suitable for casual living. The accentuation by the dome and light from the top bestow significance upon the central hall and call for devoting it to the most vital (or prestigious) function within the house. The solemn biaxial symmetry impedes casual changes, spatial rearrangements or subdivisions.

For contemporary living and working spaces, however, the nine-part plan is well suited if its centrality and hierarchy are dissolved and replaced by equivalent rooms in a square grid, which also corresponds to a more egalitarian social order. Turning certain spaces into courtyards would improve natural lighting. Individual accessibility of the rooms could be further improved by inserting a circulation spine (Fig. 11.7). Thus, the scheme would be suitable for low and medium density single or double storey housing and would allow for modular growth and the rearrangement of room uses.

(a) Panahi House, Boshroye: open central space

(b) House in Zavvare: semi-open central space

(c) Azimian House, Sabzevar: domed central space

(d) Houses in Zavvare: originally semi-open central spaces

Fig. 11.6 Types of central spaces of chahar-soffeh houses. *Sources* **a** and **c** authors, **b** and **d** Deimary (2012)

On the periphery of Sabzevar, where plots are available at moderate prices, terraced houses and low-density residential complexes are being built which mostly correspond to global standard schemes. Modernised courtyard houses could achieve the same densities and thus ensure the economic efficiency of the building projects. The currently valid zoning plans and building regulations would allow the construction of courtyard houses, but under the imperative of a forced modernisation of urban planning western building typologies were adopted without taking local housing culture and climatic conditions into account. Only informal settlements that are built without official permits usually consist of courtyard houses.

The modified chahar-soffeh house type would offer countless variations for residential buildings and combinations of living and working spaces. Only a few are shown in the Figs. 11.8 and 11.9.

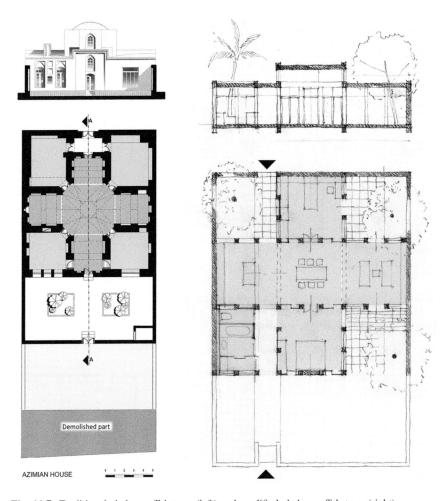

Fig. 11.7 Traditional chahar-soffeh type (left) and modified chahar-soffeh type (right)

11.3.3.2 The Linear House Type

The Baqani House (Fig. 11.10) is an example of a linear house type. It is composed of layered spatial zones: (A) The courtyard—an open space, (B) the veranda—a roofed semi-open space facing the courtyard, (C) a row of rooms connected to the veranda and to each other, (D) a potential second row of rooms in the back, only indirectly lit via the front rooms. Although layer D does not meet contemporary comfort requirements for living rooms, it could, properly dimensioned, accommodate secondary spaces such as bathrooms, toilets, storage rooms, corridors and staircases. Sanitary facilities can be designed very efficiently if they are confined to this rear zone. The poorly lit rooms in the basement which served as cool places in

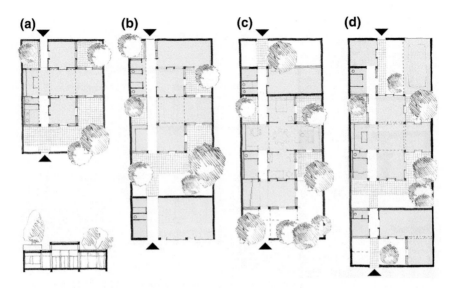

Fig. 11.8 The chahar-soffeh type revived and transformed: with circulation spine and sanitary rooms on the left side. **a** Basic scheme, **b** larger house with office or workshop, **c** with home-office, **d** with guest apartment, carport and office or workshop (Raith)

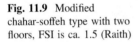

Fig. 11.9 Modified chahar-soffeh type with two floors, FSI is ca. 1.5 (Raith)

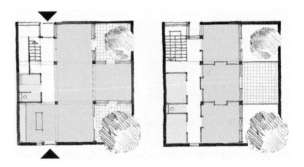

summer and storage in winter do not meet today's requirements. Geothermal cooling with heat pumps can provide this service regarding climatisation.

The clear zoning into courtyard, veranda, main spaces and secondary spaces, as well as a modular design of the main rooms, qualifies this composition for the use in terraced houses and even multistorey housing. The apartment units can be placed in a row or stacked above each other. A rear circulation zone that is directly connected to the street could provide access to the units. This would allow them to be combined or subdivided as desired, thus increasing flexibility. It would also add to the privacy of the veranda zone and the courtyard. In the case of a multistorey building

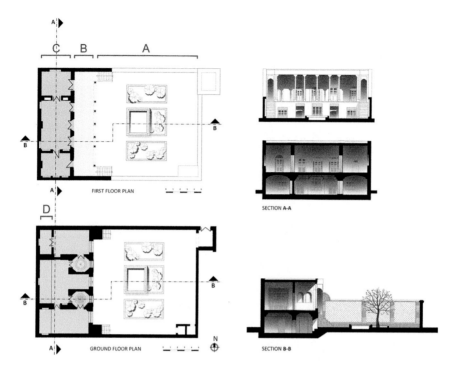

Fig. 11.10 Baqani House (Estaji 2017; Kermani-Moqaddam 2002 and the local archives of ICHTO)

the courtyard would turn into a semi-public space. Then the veranda, veiled and upgraded with various screens, could provide a kind of alternative private courtyard on each floor (Fig. 11.11).

11.3.3.3 Undesirable Typological Developments

More recent buildings from the Pahlavi period such as the Eslami house (see Fig. 11.4) are very much influenced by Western architecture with regard to the design vocabulary and the organisation of spaces. Their thermal comfort is poor due to the lack of verandas. The individual accessibility of the front rooms facing the courtyard was improved by inserting corridors on the rear side, and the lighting of the back rooms was enhanced by windows to the street thus compromising the introverted concept of the courtyard house. Altogether, this hybrid of Eastern and Western concepts does not represent a promising house type.

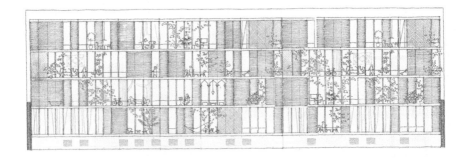

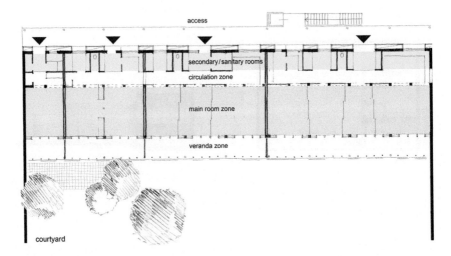

Fig. 11.11 The linear house type modified for a multistorey apartment building with distinct zones: courtyard, veranda, main rooms, circulation zone and secondary rooms. Apartments with variable room numbers (Raith)

11.4 Conclusion

Sabzevar makes a case for the following general conclusions:

- As fossil fuel becomes more expensive, it makes sense to benefit from the principles of energy-saving architecture that have been developed over generations.
- As the pace of social change increases, residential and mixed-use architecture which offers the benefits of traditional house types—neutral spaces that can be rearranged with minimal interventions—provides the best options for an adaptation to new lifestyles.
- As cities lose their cultural identity, a reinterpretation of traditional typologies by use of advanced construction methods and contemporary design vocabulary

is capable of enhancing the local character. There are no serious economic arguments against the construction of typologically revised courtyard houses; it is only up to the urban planning authorities to encourage this through appropriate zoning plans and building regulations.

It appears that in Sabzevar it is already too late to preserve the impressive architectural heritage on a larger scale, but it would at least be possible to establish continuity between past and present by reviving the outstanding achievements of traditional houses in applying them to contemporary architecture. The monuments may have disappeared, but their qualities should live on.

References

Deimary N (2012) Destruction of historical fabric of Zavvare (in Persian) http://bit.ly/2sHiOeT. Accessed 11 May 2016

Drouville G (1819) Voyage en Perse pendant les années 1812 et 1813. Paris, pp 84–85

Estaji H (2017) Flexible configuration and environmental adaptation in traditional houses of Sabzevar, Iran. Doctoral Thesis at University of Applied Arts Vienna

Estaji H, Raith K (2016) The role of Qanat and irrigation networks in the process of city formation and evolution in the Central Plateau of Iran, the case of Sabzevar. In: Arefian FF, Moeini SHI (eds) Urban change in Iran. Springer International Publishing

Keall EJ (1974) Some thoughts on the early Iwan. In: Kouymjian D (ed) Near Eastern numismatics, iconography, epigraphy, and history, studies in honor of George C. Miles. American University of Beirut, Beirut, pp 123–130

Kermani-Moqaddam M (2002) Registration report: the Baqani house (in Persian). Iran Cultural Heritage, Handcrafts and Tourism Organisation (ICHTO)

Kheirabadi M (2000) Iranian cities: formation and development. Syracuse University Press, US

Mahdavi S (2009) QAJAR DYNASTY xii. The Qajar-Period Household [Online]. Encyclopaedia Iranica. http://www.iranicaonline.org/articles/qajars-period-household. Accessed 12 Sept 2015

Mostowfi A (1997) Šarḥ-e zendegāni-e man: tārīḵ-e ejtemāʿi wa edāri-e dowra-ye Qājariya, 3 vols., Tehran, 1942, tr. Nayer Mostofi Glenn as The Administrative and Social History of the Qajar Period, 3 vols., Costa Mesa, Calif. I, 177–78

Nafisi S (2002) Be rewāyat-e Saʿid Nafisi: ḵāterāt-e siāsi, adabi, javāni, Tehran

Towhidi-Manesh G (2002) Registration report: Azimian Historical House (in Persian). Iran Cultural Heritage, Handcrafts and Tourism Organisation (ICHTO)

Chapter 12
Can Modern Heritage Construct A Sensible Cultural Identity? Iranian Oil Industries and the Practice of Place Making

Iradj Moeini and Mojtaba Badiee

Abstract The processes of urban modernisation have frequently been vilified in developing countries, and for good reasons. They are generally seen as the results of top-down decision makings and a skin-deep, out of place political will to modernise communities without having its prerequisites in place. This kind of modernisation is prone to the risks of disrupting organically and historically developed settlements and alienating people from their physical context. They are likely, as they have often been in Iran, to prioritise efficiency, facilitated traffic flows and a modernised way of development in general, over the sustainability of communities and their well-being. The question is, however, what if there is no built environment tradition in place? What if a new community is shaping around emerging forms of production in places with no established traditions of built environment? Can such new developments create a sense of attachment rather than alienation among their dwellers—unlike what is usually expected from these developments to do? This chapter looks at some of the new developments built by oil industries in their heyday in Abadan, a scarcely populated area with little recent histories of urbanism. Built to create a community from an assortment of immigrant workers, they are generally believed to defy the consensus that such top-down developments with their design ideas imported from elsewhere are incapable of creating a sense of place. Focusing on Abadan's Braim and Boverdeh, the chapter examines how, standing the test of time and still sought-after places, these developments have achieved what many of their counterparts have failed to achieve. The generally acknowledged achievements of these developments give strength to the argument that it is perhaps not so much urban modernisation per se than is the way they are viewed and implemented that results in the alienation which comes with them. In this sense, urban modernisation is not so much obstacles than are vulnerable drivers for creating cultural identities and the sense of place.

I. Moeini (✉) · M. Badiee
Shahid Beheshti University (SBU), Tehran, Iran
e-mail: h_moeini@sbu.ac.ir

M. Badiee
e-mail: m-badiee@sbu.ac.ir

Keywords Iranian oil towns · Urban modernisation · Urban alienation · Sense of place · Khuzestan · Abadan

12.1 Introduction

> The neglected … Abadan was selected as the centre of Iranian oil operations, creating a paradise in the middle of a desert by the way of residences in British style: A first experience with a new style class society formed with unified houses with small yards and red pitched roofs and new types of streets separating houses through a number only. People were supposed to assimilate into British [immigrant worker] society and their values. … Everyone in town was willing to have the experience of living in this kind of place. (Adoptions from Haghighi 2015: 32)

The story of indigenous people welcoming colonial lifestyles and its built environments is perhaps not unheard of. Nor is, however, the development of these environments into rundown areas struggling to cope with deep-rooted traditions of local life. This has been the case for Abadan's oil industry labourers' districts, where the post-war decline in local oil industries resulted in selling-off those houses and the subsequent replacement of them with new developments based on property market logics. The story goes a different way, nevertheless, with the town's 'staff' quarters, notably Braim and Bovardeh: quarters kept under the industry's control. Despite their age, they are considered by many as unmatched achievements in place making and well adaptable environments for the present-day user in Abadan. This is at odds with the common perception about such out of place developments being doomed to failure. Based on an ongoing research on the area, the present chapter is an overview of these quarters' inception, development, and their rather oblique connections with their context, followed by an investigation of how these developments have managed to work and last.

12.2 Roots: The Local Vernacular Architecture

The built environments made by tribes living in the area are not well documented. A review of better preserved built environments of other towns in Khuzestan, however, indicate well-developed building traditions. In her survey of Ahwaz and Khorramshahr, for instance, Ahari (1992) mentions the heavy masonry structures used with loadbearing walls on two sides only, with roofs mainly built with timber covered with wicker and thatch. She also mentions high-level openings in walls to facilitate ventilation. Ahari and Kasmai (1990) give a picture of a local building typology which is not fundamentally different from buildings in central Iran:

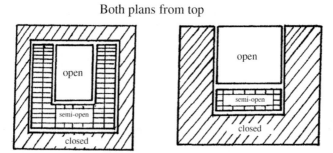

Fig. 12.1 General spatial organisation patterns in Khorramshahr and Abadan (Ahari 1992)

introvert buildings with enclosed or occasionally semi-enclosed courtyards. There are, however, differences including the extent of shaded semi-open spaces, and habitable roofs protected with lattices (Fig. 12.1).

12.3 Roots: The English Garden City

Imported by British designers, the designs of many early urban areas and residential districts for Iranian oil industries are said to be influenced by the English Garden City concepts, notably for their low density and the significant amount of greenery. With many mismatches between socio-geographical contexts of the birthplace of the concept and that of Iranian oil-rich areas, it would not be unexpected to find such imports unworkable. Many of the factors mentioned as those invoking interest in the idea elsewhere, for example 'the progressive rejection of the big city, the desire for small town living and working, the search for real involvement on common affairs, and … the adherence to a new "green" life style' (Ward 1992: 1) can be said to have played no direct role in Iranian oil towns. A desire for casting new forms of social life, however, might be seen as shared between the two contexts, with the original society seeking it as an alternative to rapidly industrialising cities and the Iranian oil town a starting point from scratch shaping previously non-existent communities out of an assortment of immigrant workers from a wide range of backgrounds and skills. Among features Aalen (1992: 29–30) mentions about Howard's Garden City are those of small, thoroughly planned towns with each town balanced both socially and economically, 'accommodating all classes and providing a range of employment in primary, secondary and tertiary activities':

> The city is conceived, like many utopias, on a circular basis and there is a clear zoning system within it. Service buildings and public buildings are at the centre with a belt of residential land around them and the railway and factories are on the perimeter. Public gardens, parks and tree-lined avenues are prominent features. (30)

This is, on one hand, a far cry from local traditions of built environments in Khuzestan, but it also corresponds with the emerging communities in the area's oil industry workers' districts: the coming together of modern-day workers settled close to their workplaces in places with no remarkable histories of built environment (Fig. 12.2).

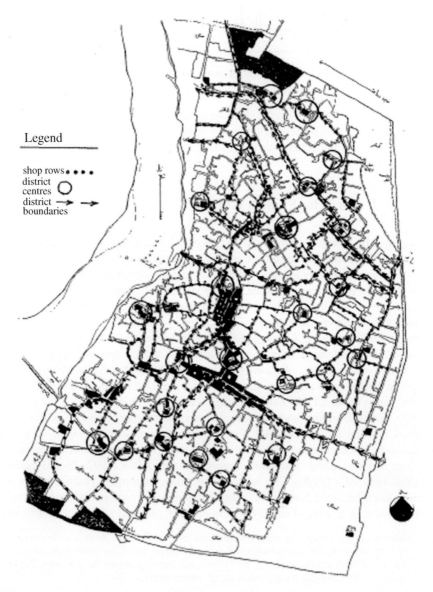

Fig. 12.2 A typical traditional town in Khuzestan (Rahimieh and Roboobi 1974, adopted here from Ahari 1992)

12.4 Roots: The Emergence of Oil Industries in Iran

The history of oil industry settlements in Abadan goes back to early twentieth century, when a new town was formed on a sparsely populated island on the estuary between Karun and Euphrates-Tigris. It consisted of the main refinery area, complete with ancillary buildings, and early bungalows built using local construction techniques and materials. The early houses, however, soon gave way to new ones closer in design to British style bungalows and semi-detached houses. There was, from the beginning clear separation between different workers' socio-economic classes and ethnicities, which resulted in having boundary lines between different quarters of the town then and even today. In the case of senior and intermediate staff quarters, namely Braim and Bovardeh (Bawarda), not only have the boundary lines continued to be recognisable until today, but also significant proportions of original buildings have survived and still used by the workers' classes they had originally been designed for (Fig. 12.3).

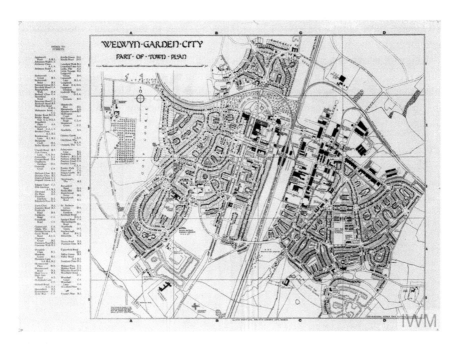

Fig. 12.3 Welwyn Garden City. *Source* http://www.iwm.org.uk/collections/item/object/205133059

12.5 Abadan: The Structure of the Master Plan and Its Subsequent Developments

In his chapter on the social aspects of Abadan's inception and development, Haghighi (2015) offers a rather grim view of the town's development strategies, pointing, inter alia, to the above mentioned segregationist approach which draws unsurpassable boundaries not only between British, non-British and Iranian workers but also between senior staff and plain labourers. This, in his view, goes so far as to completely ignoring climatic control considerations in labourers' houses. There is in his work also a criticism of enforcing a modern lifestyle of nuclear, isolated families who lose their connection with their roots both through their inability to form extended family settings, and to have at least domestic-scale in-house farming. The divide between quarters is also criticised for being motivated by crowd control and therefore possible rebellion containment strategies imposed to safeguard a trouble-free running of the industry. As he observes, there is a wealth of public amenities in senior staff quarters, but that they hardly form monuments to help reinforce a sense of place. Besides its thoroughly managed quarters, he observes, emerges quarters emerged more spontaneously which, substandard as they might have been, have enjoyed a vividness perhaps not seen in the quarters more directly developed by the industry (Fig. 12.4).

Relevant as the observations and criticisms like Haghighi's are, they represent a much recurring model for modern top-down urban developments, where colonial or other ideals of efficiency, modernisation and their associated social transformation are imposed on historically evolved communities ignoring genius loci and the

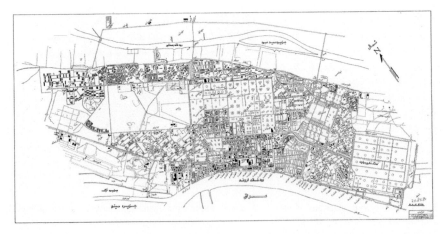

Fig. 12.4 An old map of Abadan showing refinery and other clearly demarcate quarters (NIOC)

deep-rooted built environment insights on the way. In comes, as any developing country has experienced, vehicle-friendly networks of roads, commodification of properties and new standards of public services. Out goes, however, lifestyles in harmony with nature and community, the sense of place and consistent growth. In Khuzestan's case, even as considerate and sympathetic a scheme as Shushtar New Town by Kamran Diba has suffered incompletion and damages to the original scheme due to changes in circumstances, turning it into a no-go zone for many (Shirazi 2016, pp. 135–137).

This is, however, not necessarily the case with Abadan, notably with Braim and Bovardeh as both quarters have more or less stood the test of time and are still under a similar management system—with this management consistency and the unchanged company ownership admittedly playing parts in their stability—and used by oil industry workers. What follows is a look into how these quarters have managed to buck the trend, and whether there are lessons to learn from their survival and thriving.

12.6 Abadan: The Architecture of Oil Industry Settlements

Throughout their ups and downs, Company-managed oil industry settlements appear to have kept their focus on inhabitation rather than mere housing, and thereby succeeding in place making. As early as 1909, states Crinson (1997), it was recognised that everything needs to be imported into Abadan in order to develop it. This was not, however, limited to building materials and technologies, but also expanded to building types and urban design. According to Rostampour the first Western building types made its way into the region, with the peripheral massing around a courtyard giving way to a central massing surrounded by greenery, and with sanitary services and the kitchen relocated into the building. These types were mainly for higher ranks of staff, but there was in turn a hierarchy between them, on top of their distinctions from manual labourers (2016: 87). Braim, for example, housed five zones roughly corresponding with the seniority of their inhabitants. The houses, particularly those of intermediate and senior staff, were divided into 'main' habitants' and servants' zones. According to Crinson (ibid.) electrical appliances such as fans were used for ventilation. More importantly, green spaces were introduced both to moderate the harsh climatic conditions of the region and to resemble the British atmosphere for expatriates. These settlements were initially designed to inhabit isolated communities of British and other Western workers enjoying their own exclusive facilities, and without much need to be in touch with local people (Rostampour 2016, p. 91).

This isolationist approach is in line with the zoning strategies in Abadan master plan which, according to Ardalan (2010/2016) is in turn in line with modernist urban design strategies. A key figure in this period is the British architect Wilson (Crinson 1997) who was behind the implementation of Garden City and zoning ideas. He also developed detached types into semi-detached types (Ardalan 2010/2016), which was established as one of the most widely used types. The type somehow corresponded with the region's climate by eliminating exposure to sun on one side of each unit (usually east and west facing sides). This was followed later by adding garages to the detached side and thereby eliminating sun exposure for another side. The variations in designs keeping settlements away from an imposed uniformity (Rostampour 2016: 95), and also an attempt by Wilson to make use of elements of vernacular architecture (Crinson) such as flat roofs, recessed clerestory windows and occasionally playful brickworks are among reasons cited for these settlements' success. According to Rostampour Wilson's 1934 design for Bovardeh, took a step further by paying more attention to site and open spaces—no longer merely leftover spaces—and designing as many as 27 types for the southern part only. The variations in typologies also indicate the presence of an experimentalist attitude in design. There are instances of both flat roofs and additional ventilated pitched roofs acting as shading devices to the main roof. There are also experiments with high-level ventilation, external window blinds, deep verandas, and regionalist references, particularly the use of brickworks and ceramics, not to mention experiments with solid-void organisations in plots. This degree of variety, however, was not followed in later developments (Ardalan), resulting in more uniform complexes within more rigid urban grids, and thereby losing urban design qualities associated with the likes of Braim and Bovardeh.

Another important development in architectural design, though not unparalleled elsewhere in Iran, is the new houses' ability to accommodate modern appliances and furniture (Rostampour 2016, p. 97) something with profound implications on local inhabitants' ways of organising interiors, and thereby moving towards more functionally determined laying out of spaces. Also important was the introduction of piped water services (ibid.) which, again, contributed in another important aspect of these houses, namely, creation and maintenance of green spaces.

In terms of urban design Braim and Bovardeh are both low density quarters with plenty of open spaces and wide streets. This does not help shaping communities other than in community centres such as clubs and sports facilities. The streets, however, are carefully designed with sizeable drainage canals, and designated pedestrian and greenery lanes on both sides. Generously sized open spaces and alleyways providing access to service parts of houses are also in abundance. Many of the houses located on street corners or urban nodes enjoy special design, notably the houses designed with 'towers' at one of Bovardeh's entryways (Figs. 12.5, 12.6 and 12.7).

a. A street in Braim; the curved lines, the greenery, and designated pavements are among important features

b. Braim: Most houses are well hidden behind hedges

c. Braim: communal open spaces

d. Braim: open front gardens in semi-detached houses

e. Braim: designated service blocks' access streets

f. Braim: semi-detached servants' blocks

Fig. 12.5 a, b, c, d, e, f Present spatial characteristics of open spaces and houses in Braim and Bovardeh, Photos by authors

a. Braim: deep verandas

b. Braim: an architecture of brick ornaments and lattices, verandas and ceiling-level ventilation

c. Braim; external blinds and ventilated secondary pitched roof

d. Bovardeh; feature tower, deep shading device and ornamented faÅade, one of the entryways

e. Braim; ventilated secondary pitched roof

f. Bovardeh; row houses with service blocks facing back street

Fig. 12.6 a, b, c, d, e, f Present design elements of housing typologies in Braim and Bovardeh. Photos by authors

Fig. 12.7 Bovardeh; roundabout building type, Photo by authors

12.7 Braim, Bovardeh and the Rest

Figure 12.8 shows the master plan of Braim. Originally designed to house the Oil Company's senior British staff, it is comprised of detached and semi-detached houses with a variety of sizes and massing, as well as communal open spaces and amenities. Whilst many units are facing north and south the flexibility of the access

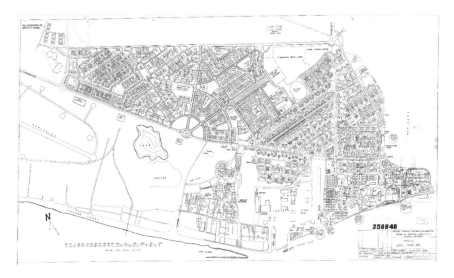

Fig. 12.8 Site plan of Braim quarter. *Source* NIOC

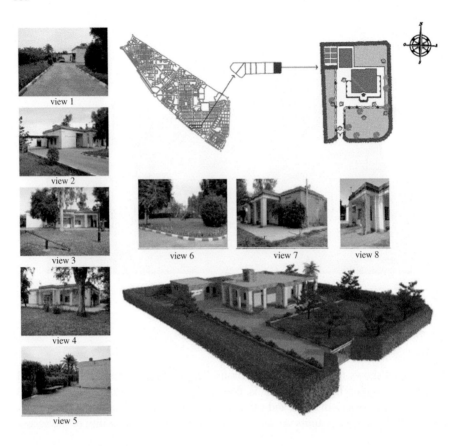

Fig. 12.9 A detached house in Braim with a separate service block. *Source* Authors

network places the quarter far from rigid and closer to those of Garden Cities. As mentioned earlier, and despite being built in a relatively short period of time between 1925 and 1940, there is a wide range of varieties, but most types are single storey brick buildings with flat roofs, means of ventilation through ceiling and verandas.

Built for medium rank staff Bovardeh (Fig. 12.9) is divided into two parts with less greenery and relatively higher density. But the spacing between units still leaves plenty of open space, and there is still a wide variety of types, notably types designed in urban nodes. The references to regional architecture are visible in bands of ceramic ornaments on top of ornamental brickwork (Figs. 12.10, 12.11 and 12.12).

There are other quarters made by the Oil Company for their staff and labourers, which have changed hands following the post-war decline in Abadan's oil

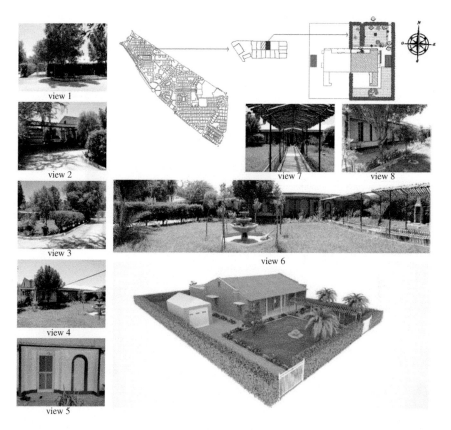

Fig. 12.10 A semi-detached house in Braim with secondary ventilated pitched roof, garage and service block attached to the building. *Source* Authors

industries. Lacking the Company's management, many of these houses have undergone profound alterations and substandard reconstruction, and left in a state of dilapidation. There are, however, some original elements left, which indicate parallels with senior staff housing design strategies. These include the presence of a front, open-to-street garden, controlled daylighting through feature windows with limited openings, and ceiling-level natural ventilation. The blocking of a thorough view to street through brick curtains as well as a direct rubbish disposal system which eliminates the need for residents to appear at the door to dispose rubbish, however, are among features not repeated in senior staff quarters (Fig. 12.13).

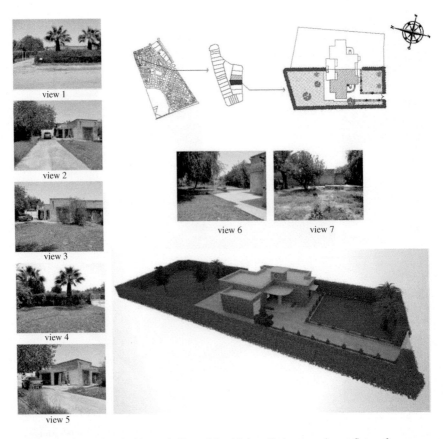

Fig. 12.11 A semi-detached house in Bovardeh with 'ventilation tower' over flat roof: a common feature in houses with flat roofs. *Source* Authors

12.8 Conclusion: Modernisation Through Built Environment; A Natural Failure?

In his research on the problem of identity and identification in Khuzestan's oil towns Rostampour (2016: 64) observes that like many other instances the new settlements created by, or resulted from developing oil industries were based on entirely different modes of economic productivity: 'modern towns based on social engineering and new economic relations based on industrial economy'. Ehsani (1999/2016: 25) states that the Oil Company architecture aimed not only at mass production of cheap, durable housing, but also at a modernisation of the traditional family structure. According to Rostampour (80):

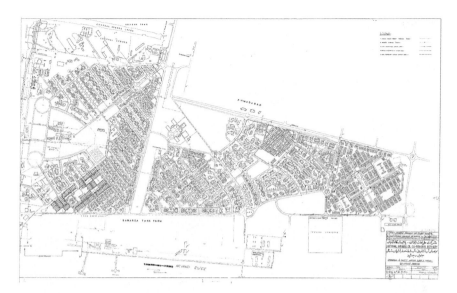

Fig. 12.12 Site plan of Bovardeh quarter. *Source* NIOC

> This architecture neither provided adequate space to accommodate extended families—the local norm—nor allowed a productive, economic use of home. Instead, it offered new forms of sanitary services such as kitchens and bathrooms, as well as semi-fixed elements such as beds and desks … leading to the formation of a different identity to that of families' previous identities.

Interestingly, he also observes that the success of these oil industry settlements in constructing new identities remained unmatched by others in the province, including the architecturally important New Shushtar by Kamran Diba built for sugarcane industries (84). In his attempt to investigate the reasons for this success, he points (93–95) to items such as the gradual shift of their design, particularly in their middle period (1934–1939) towards some kind of regionalist sensitivity—for example in the use of local materials and labour, an attention to the location, the use of ornaments and brickworks akin to local architecture, the provision of new alternatives for community relations through clubs, swimming pools, cinemas and schools, and also a sense of satisfaction and pride which, he argues, precedes the quality of facilities. To those, one must add other reasons discussed less explicitly by Rostampour. For example, although there is evidence of a deep history of settlements in what is now Abadan, it appears to be, at the time of the emergence of oil industries, more the site of some tribal settlements, and nothing in scale and development like, say, Dezful. This led to the formation of a new community almost entirely formed by migrants, albeit from very different backgrounds, with no recognised genius loci or consensus about modes of life. A tabula rasa, something modern architecture and urbanism is usually accused of assuming, is probably not as out of place here as it might have been in more historic contexts. The

a. Interventions in labourers' housing types

b. Labourers' type with secondary ventilated roof

c. Lattice parapet in labourers' houses providing night time habitable space at roof

d. Brick screens in labourers' quarters blocking direct view into streets

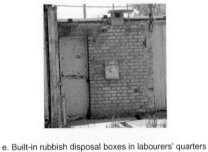
e. Built-in rubbish disposal boxes in labourers' quarters

f. Secondary ventilated roof and feature windows in labourers' quarters

g. Rundown streets in labourers' quarters

Fig. 12.13 a, b, c, d, e, f, g Present design characteristics of labourers' quarter. *Source* Photos by authors

philanthropic ideas of Garden Cities could thus be implied with minimal risks of failure caused by its mismatch with local historic, if not climatic paradigms. These settlements actually appear to have gone further by moderating the harsh local climate through introducing significant greeneries, as well as preserving the locally desirable sense of privacy, not through impenetrable walls—as done in labourers' quarters—but through defining softer, more climate-friendly green boundaries. Furthermore, controversial as it might be, the segregationist strategies used in master planning, created communities of people who find more in common with their neighbours carrier-wise despite their possibly different ethnic backgrounds.

The successful examples of oil industry settlements such as Braim and Bovardeh belies the claim that the move away from local built environment traditions is doomed to fail, and that the capability of built environments to transform societies will always result in uprooting people and maligned communities. They have been shaped at the time of profound changes in models of economic productivity and community formation, and have at least corresponded to these emerging forms, if not engineered it towards new directions. Moreover, there have been a range of experiments to 'regionalise' buildings through climatic solutions and construction and ornamental reference, which have managed to maintain a sense of place despite entirely overhauling lifestyles and community formations. These are, of course, helped by having a functioning, consistent system of ownership and management to, by and large stop the quarters from turning into rundown areas. One might argue that such subsidised systems are no longer viable. Nor, it can be argued, are the generous open and green spaces are affordable any longer in a world quickly running out of resources. There, are, however, many lessons in Abadan's oil industries built environment achievements as to how to create a sense of place away from the ruthless logics of market-laden built environments, in a context of politically charged, top-down decision makings in a modernising society.

References

Aalen FHA (1992) 'English Origins' in Ward
Abadan map https://ajammc.com/wp-content/uploads/2015/02/Abadan-Map-Hi-Res-1024x389.jpg
Abadan kmap http://www.adababadan.com/map05.html
Ahari Z (1992) The house building pattern in Khuzestan. Iranian Building and Housing Research Centre (in Persian)
Ardalan N (2016) Email interview by Kaveh Rostampour, 2010 (quoted here from Rostampour 2016)
Crinson M (1997) Abadan: planning and architecture under the Anglo-Iranian Oil Company. In: Planning perspectives, vol 12, Issue 3. History of Art Department, University of Manchester, https://abadancm.com/2016/06/15/abadan-planning-and-architecture-under-the-anglo-iranian-oil-company/
Ehsani K (2016) Modernisation and social engineering in Khuzestan oil towns: A Glance at the Abadan and Masjed-i-Soleiman experience. In: Goft-o-goo, No 54 (in Persian, quoted here from Rostampour 2016)

Haghighi S (2015) Sociology of life and building in company town. Int Res J Appl Basic Sci 9(4)
Kasmai M (1990) Khuzestan Climate and Architecture—Khorramshahr. Dad (in Persian)
Rahimieh F, Roboobi M (1974) An introduction to towns and houses of Iran. Tehran University Student Association, Hot and Semi-humid Climate
Rostampour K (2016) Explaining the role of interaction of housing and regional characteristics in residential architecture identity; case study: oilfield dwellings. PhD Thesis, University of Tehran
Shirazi MR (2016) From Utopia to Dystopia: Shushtar-e-No, endeavour towards paradigmatic shift. In Arefian F, Moeini I (eds) Urban change in Iran; stories of rooted histories and ever-accelerating developments. Springer
Ward SV (ed) (1992) The Garden city; past present and future. E & FN Spon

Chapter 13
Evaluation of the Prospective Role of Affordable Housing in Regeneration of Historical Districts of Iranian Cities to Alleviate Socio-spatial Segregation

Alireza Vaziri Zadeh

Abstract In many countries, the poorly constructed and socially deprived areas in the historic part of the city—known as declining areas—have been the subject of a variety of intervention approaches. Housing-led approach in regeneration aims to tackle the failure of commodified housing market, meet housing needs of deprived households, mitigate the social deprivation and poverty and improve the area's built environment. This chapter investigates how housing-led approach in regeneration programmes has the ability to meet the low-income housing needs decently and tackle social exclusion; and how this approach might contribute in restructuring cities. During the last decades in Tehran, declining areas have been subjected to many urban regeneration programmes. Here contemporary programmes are categorised into two generations. These programmes—particularly the second generation—not only took the redevelopment of declining areas in historic zones into account, but also looked at the provision of (affordable) housing, to meet the overall accelerating housing needs caused by rapidly growing urbanisation, demographic change and unaffordability of houses in the city. However, success degrees are disputable, particularly in terms of the ways in which they have restructured the socio-spatial pattern of the city and whether the targeted areas moved towards social inclusion or not. Evaluating advantages and disadvantages of contemporary regeneration in Tehran as well as some other international experiences, the chapter argues that in order for affordable housing-led approach to regeneration of declining areas in the historic zones to be successful five key points need to be considered: (a) a developed monetary system has to secure the financial aspects of programmes, (b) to alleviate the negative impact of interventions, the changes have to take place in long period, (c) the participation of private sector is crucial, albeit with reservations, (d) each programme requires its own innovative methods to encourage participation of people, (e) on top of spatial development, the empowerment of deprived households through various methods is required.

A. Vaziri Zadeh (✉)
Faculty of Engineering Science, Dept. of Architecture, Planning & Development,
University of Leuven, Leuven, Belgium
e-mail: alireza.vazirizadeh@asro.kuleuven.be

© Springer Nature Switzerland AG 2020
F. F. Arefian and S. H. I. Moeini (eds.), *Urban Heritage Along the Silk Roads*,
The Urban Book Series, https://doi.org/10.1007/978-3-030-22762-3_13

Keywords Regeneration · Historic districts · Declining areas · Housing-led approach · Iranian housing system · Welfare system

13.1 Introduction

This chapter aims to explain the extent to which the housing-led approach in regeneration—particularly the provision of affordable housing—can meet low-income housing needs, and how this approach engages in restructuring of the cities? During the last decades, and in many countries, the poorly constructed and weakly facilitated areas in the inner parts of cities—known as declining areas— have been the subject of a variety of interventions. In Iran, and particularly during the last two decades, there is a new wave of attention to urban redevelopment programmes that on one hand pays attention to the regeneration of historic declining areas to decrease the area's vulnerability against natural disasters, and on the other hand looks at the provision of affordable housing to meet the overall accelerating housing needs that stem from fast growing urbanisation, demographic change and unaffordability of housing market.[1]

To tackle the failure of housing market in declining areas and to meet housing demands of the deprived households, the government (as the main agent of development) has initiated some area-based initiatives to regenerate historic areas (with a history of social exclusion and massive urban intra-migration and segregation). On top of re-envisioning their built environment, the regeneration programmes promised to provide affordable housing for low-income households, mitigate deprivation and poverty, and balance the housing market. However, the degrees to which these initiatives have restructured the socio-spatial configuration of the areas to alleviate exclusion[2] levels, is disputable.

In the following sections, and based on contemporary international practices, the chapter first reviews the advantages and disadvantages of housing-led regeneration approach in recent regeneration programmes (mostly in European countries).

---

[1]Affordability of housing is the subject of wide-ranging interpretation. Arguably, the relation of house prices to the household income is the most well-known measure for affordability. However, the affordability of housing interrelated with more general issue of macroeconomic performance and function of the housing market (Roberts et al. 2016).

[2]There is a direct relationship between key determinates of social exclusion and poverty, i.e. the lack of ability to find employment, pursue education; and the problems of environment and health related to the housing (MacInnes et al. 2013). Social Exclusion is defined as *"The increasing spatial concentration of the disadvantaged [which] has resulted in the isolation of many individuals and households from mainstream social and economic activities"* (Mcgregor and McConnachie 1995, p. 1), resulting in exclusion of individuals from labour market.

Then, it focuses on contemporary history of regeneration in Tehran. Reviewing the practice during the last two decades, the chapter argues that housing-led approach to the regeneration of declining areas can be successful if some key points are considered.

13.2 General Perspective

Interventions in the built environment and spatial rearrangement have been carried out in regeneration schemes to counter urban decline and poverty concentration in deprived areas. Such deprived areas usually contain vulnerable households, and commonly stigmatised as 'declining' (that makes intervention eligible). The population living in those areas are on average under-ranked in terms of some social indicators, and suffer poor built environment conditions such as overcrowded housing, lack of services and weak access to transportation (Pugh 1997; Smith 1999; Vaughan and Arbaci 2007).

13.2.1 Regeneration (Area-Based Programmes)

Urban regeneration is a generic elastic term that covers different strategies of intervention, and it can be vaguely defined (Jones and Evans 2008).[3] Yet, regeneration in this chapter refers to the Porter and Shaw's (2009, p. 2) definition. They believe that regeneration means: 'reinvestment in a place after a period of disinvestment'.

Regeneration initiatives have been put forward to confront urban poverty and inequality (particularly in many European cities after the 1990s). The main aim of most regeneration programmes is the revitalisation of declining communities; whether it is going to face with spatial, economic or social issues (or all of them) (McCarthy and Pollock 1997). The initiatives implicitly intend to respond to poverty and disadvantage to promote social inclusion and enhance the quality of life. The cores of the cities mostly including and/or surrounded by poor communities are the most attractive places for regeneration projects (Drakakis-Smith 1981; Mcgregor and McConnachie 1995).

---

[3]In the literature, different terms are used to refer to various strategies. For instance, The programmes that aim to reduce the social and benefit low-income people are tended to be called 'regeneration', while the programmes focusing on re-imaging of the cities through massive destruction of the areas to attract capital and middle incomes are usually called 'urban renewal' or 'urban renaissance' (Tallon 2009).

13.2.2 Housing-Led Approach

It has been revealed that other approaches, for instance property-led regeneration, not only could not tackle the problem of poverty and exclusion, but also they led to considerable gentrification in cities (argued in detail in: Porter and Shaw 2009; Tallon 2009). Therefore, some conclude that those regeneration projects which just focus on development of property are inadequate in themselves (Healey et al. 1992; Imrie et al. 2009; McCarthy and Pollock 1997).

Given that multiple deprivations are linked to poor housing, it can be argued that the issues of housing and socio-economic exclusion are interlinked. Improvement in housing and environmental conditions is the prerequisite to improving social inclusion and breaking the cycle of poverty and deprivation (Keivani and Werna 2001; Pugh 1997, 2001).

Housing-led approaches have pursued different objectives in different contexts. In European practices, this approach is not supposed to meet the needs of housing market, but housing development was stimulated by a modern agenda to boost economic regeneration and attract skilled people to low-income neighbourhoods. For instance, the housing-led regeneration defined in 'Going for Growth'[4] seeks to improve economy of deprived communities through introducing more affluent populations. It applies an approach to redirect existing housing market with a significantly different market, not only in terms of the quality of housing and inhabitants, but also in tenure terms (Cameron 2003).

The idea which is encapsulated in the term 'housing-led regeneration' is applied primarily as a means for improving living conditions of communities, and an important element to stimulate the wider economy of the city. Regeneration programmes, per se, has an effect on competitiveness of the local urban economy and growth of property markets. However, the programme which applies housing-led approach to provide affordable housing, can improve the quality of life, the well-being of residents, and social inclusion within the targeted areas (Ennis 2016; Scanlon and Whitehead 2008; World Bank 1994; Cameron 2006).

13.2.3 Advantages and Disadvantages of the Approach

According to Porter and Shaw (2009), the main question in evaluating the area-based regeneration is who benefits from implementation of these programmes; or, in other words, the question should be asked is not *whether* the programmes were successful or not, but *how* and *for whom* (Also in: Arbaci and Tapada-Berteli 2012).

---

[4]"The city-wide regeneration strategy known as 'Going for Growth' adopted by Newcastle City Council includes proposals for large-scale redevelopment of low-income, low-demand housing neighbourhoods..." (Cameron 2003, p. 2367).

The success of regeneration programmes is measured primarily by the rise in land values. Therefore, regenerations generally deemed successful in improving built and natural environments have a dark side. Increasing the land value directly affects lower income people and can cause displacement and exclusion.

After long-term implementation of these initiatives, housing-led intervention schemes are dealing with unforeseen urban socio-economic issues. The assessments of the programmes in terms of their social impact, both in the Global North and South, have raised many controversial debates, particularly about gentrification and displacement of indigenous inhabitants (Arbaci and Tapada-Berteli 2012; Cameron 2006) .

The evaluation of the programmes has revealed their inability to benefit the low-income people. Cameron (2006, p. 14) argues that despite the rhetoric of the positive effects of these programmes in combating the 'social exclusion' of the poor, very little benefit from housing-led regeneration goes to lower income people. The interventions not only demolished housings of the poor households, but also violently destroyed social networks, cultural forms of life, long-established mechanisms for economic survival and tenure structure. Under these circumstances, the most marginalised people from the economic mainstream (the low-income tenants) were the most vulnerable and suffering people (also in: Gutiérrez Romero 2009).

The poverty that has been the subject of the area-based programmes is the outcome of socio-economic inequality in the society, and only slightly rooted in the living conditions of declining areas. By drawing a boundary for initiatives, and not facing issues of 'poverty' and 'deprivation' in a broader context, the benefits of initiatives moved towards landlords, developers and investors of the areas (Lupton and Tunstall 2008, pp. 114–115). Wealth is certainly generated, but as Porter and Shaw (2009, p. 3) argue the question is wealth for who. While the 'trickle-down' effects of the programmes have not worked, the wealth is concentrated in fewer hands, and poor communities and their surroundings have gained little opportunities. All these unforeseen impacts have emerged because the programme did not tackle the main source of deprivation; conversely it worsened the issue of affordability for vulnerable households (Arbaci and Tapada-Berteli 2012; Cameron and Doling 1994; Tallon 2009).

However, by substituting new and more affluent people, housing-led regeneration projects wipe out issues of locality rather than facing with the economic and social problems of existing communities. What housing-led regeneration, implicitly or explicitly, promised and moved towards is a kind of engineered gentrification; i.e. a considerable portion of low-income households have to move and be replaced by higher income people (Robertson 1998; Tallon 2009).

Despite many disadvantages mentioned regarding housing-led programmes, there is evidence that if there were grass-root actions in initiatives, more equitable outcomes would be achieved (argued in detail in: Porter and Shaw 2009). Whenever there was a commitment to socially progressive outcomes to address both physical and social issues and valuing different kinds of 'reinvestment' (particularly sufficient provision of affordable housing), more positive outcomes are expected.

Based on an evaluation of initiatives during 1990s, Tunstall and Lupton (2003) conclude that the effectiveness of the programmes for the poor depends on whether the right group is targeted. According to them, the programmes have targeted the poor more accurately than claimed by opponents.

Lawless (2012) supports housing-led programmes in the UK. First, he argues that in a neo-liberal sociopolitical context, 'it is better than nothing' (p. 324). Second, there is considerable positive achievements in relation to mental health (p. 325), as residents are more appreciative of their local areas and environments. Third, there are some individual gains regarding education and jobs that could not be captured by household surveys. He states that there is evidence that many individuals benefited from interventions, even such impacts are difficult to measure (p. 313). The conflicts emerge when the explicit political assumptions of the programmes do not always balance with the local residents' wish (Lawless et al. 2009; Lawless 2006; Parry et al. 2004).

13.3 Housing-Led Regeneration in Tehran

Since early 1990s, and after the eight-year war and stabilisation of the new government, the implementation of regeneration programmes has been accelerated in all Iranian cities, particularly in Tehran, the largest and the most populated city in country. Compared with contemporary European practices, the targets and outcomes of the programmes in Iran are different from various perspectives.

13.3.1 Regeneration; Motivations and Objectives

To scrutinise regeneration practices in a society, it is crucial to understand how the concept of 'declining' is interpreted in that society; how an area is labelled as declining; what is the main reason(s) for declining; and which kind of issue(s) is associated with that place, how that phenomenon is recognised as an issue, and ultimately why it is necessary to intervene in an area. Accordingly, the motivation and objectives of intervention in an area can be manifested, and consequently, the approach to the intervention (regeneration, reconstruction, renovation and others) will be unravelled.

Contrary to the cities in the Global North where the deindustrialisation and shrinking population (low housing demand) led to a decline of the cities, in Iran, and particularly in Tehran, the rapid development and the significant expansion of new residential areas (modern neighbourhoods that are well facilitated with infrastructure) has resulted in a long-term neglect and disinvestment of old parts of cities, and has entrapped those areas in a cycle of decline (Madanipour 2006;

Sepehrdoust 2012). However, disinvestment per se did not motivate intervention in urban areas. The massive earthquakes of 1990 and 2004 were turning points that disclosed vulnerability of old parts of Iranian cities against disasters (and their consequent sociopolitical crises), and arouse motivation for reconstruction and regeneration of cities.

Moreover, since early 1990s, the government gave up the policy of massive land allocation and expansion of the cities (Azizi 1998). Thus, the value of land increased rapidly and subsequently the declining areas were reassessed as opportunities for prospective profit-making (rent-speculation) development.

Declining areas were assessed as problematic, and that legitimised interventions. As interventions are expensive and might lead to some social instability, a strong discourse has to support and legitimise regeneration programmes. Thus, some issues would have to be attributed to declining areas in order to demonise them as serious threats to cities' livability (particularly when a top-down approach was at work). Accordingly, declining areas were seen to have the following problems: (Sazman-e-Nosazi-e-Shahr-e-Tehran 2015):

- These areas have lost their comparative advantages. Private developers are not interested in investment in these areas and landlords rarely refurbish their houses. Old buildings, due to long-term negligence, are highly vulnerable against disasters especially earthquake. Small plot sizes, narrow streets, and the lack of accessibility by the vehicles, make any development very difficult.
- Given the fact that the older parts of cities have not been sufficiently served by urban infrastructures, many affluent households have gradually moved out and the housing market of these areas have greatly declined. Therefore, those areas attract very low-income households. The concentration of poverty is the source of many social offences that disturbs urban liveability. After a long-term negligence, these neighbourhoods have fallen in a cycle of deprivation and social exclusion.

13.3.2 Housing in Regeneration Practices in Tehran

The regeneration programmes during the last 25 years, in approach terms, can be classified into two generations. The first generation (approximately between 1994 and 2009) were was supported by a political will for rapid (economic) growth. The dominant projects of this generation are '*Navvab*' and '*Imam-Ali*'. Some of the characteristics of the first generation are:

- The dominance of the property-led approach
- The implementation of mega-scale projects
- Massive intervention and construction (highly ambitious targets)

- Enormous eviction and dislocation of local people
- Insufficient funding for realisation of projects
- Very challenging outcomes (evaluated as failure)

Although in both projects considerable residential area were developed, the programme did not target the lower incomes. This was particularly the case with the former, which was designed to inhabit middle classes (engineered gentrification). Yet, due to the failure of the project and the subsequent decrease in property values, the intended objective could not be met, and after a decade the new low-income households inhabited the area. These programmes changed the social structure of the areas (Etemad 2012). For instance, during the implementation of '*Imam-Ali*' project, about 50,000 people were dislocated.

The 2004 earthquake in Bam is a turning point in the history of regeneration in Iran. The necessity of regeneration of vulnerable areas in short-term and the failure of previous practices made decision-makers to re-evaluate their approach. In 2006, the 'Supreme Council of Urban Development and Architecture' recognised the vulnerable and declining areas in cities. To recognise an area as declining, three measures were considered: unstable and unsafe structures (incompliant with building codes), narrow streets (less than 6 m wide), and small plots (less than 100 m^2). The co-presence of these three characteristics qualified an area as declining and subject to intervention. Based on these characteristics, 14,792 ha of urban areas in Tehran were recognised as declining, with 3268 ha subjected to direct intervention. The measures used focused only on built environment ignoring many social and economic aspects (including those related to housing). Moreover, by setting a boundary around particular areas, the adjoining neighbourhoods were ignored, even if they were also declining.

Since 2008(9), many area-based programmes (about 50, for a total area of 2000 ha) have been initiated to regenerate declining areas in Tehran. The different approach adopted in this second generation was stimulated by the failure of previous practices (highly ambitious direct intervention), and the lack of sufficient funds. The new policy is based on indirect intervention to encourage local people to renew their vulnerable houses with the participation of private developers. To motivate both parties many benefits are offered to both the landowner and the developer, including deregulation (flexibility in land-use and urban density), reduction of bureaucratic procedures, tax relief, low-interest loans and even limited financial subsidies. Despite its successes in directing investment towards declining areas and the supply of new housing stock in recent years, the programme has only benefited landowners and private developers and the affordability of new houses is very arguable. Moreover, the deep recession in housing market since 2012 cast doubts on the future of such programmes (Kamrava 2010; Rezayee 2009).

To avoid changing the social structure of programme areas, the Municipality (the main agent for regeneration) initiated another programme (called '*home-for-home*') to provide housing for households who have inevitably lost their properties (for instance to make public open spaces). Many affordable housings were provided (30,000 units) for local low-income people through this programme. The new housings were given as a replacement for their nearby demolished home. This policy has proved successful and reduced dislocation. However, funding shortages was again problematic.

Some characteristics of the second generation of regenerations (since 2005) in Tehran are:

- Focusing more on area-based housing-led approach
- Noticing mostly Paying more attention to planning than direct project-based interventions
- The provision of some affordable housing for low-income households
- Preventing dislocation of local people (to some extent)
- The increased engagement of people in programmes

13.4 The Prospect of the Housing-Led Approach

How is it possible to improve the efficiency of (affordable) housing-led programmes? Based on its reviews of highlights and challenges of contemporary regeneration in Tehran and elsewhere, this chapter extracts important considerations to take into account for future housing-led regeneration:

(a) a developed monetary system has to secure the financial aspects of programmes to avoid periodical deficit, rent-seeking and corruption:

Some aspects of Iranian financing systems have not been developed in line with global practice in developed countries. For example, (un)secured loans, mortgages, credit facilities, lines of credit or corporate bonds are so restricted that they constrain innovative and advanced investment in regeneration programmes. Moreover, there is evidence of rent-seeking and corruption in many development programmes among managing agencies who have access to information and funds. Accordingly —as evident in many cases—the achievement of programme targets faces grievous challenges, such as long accomplishment times and the diversion of benefits to the wrong people (Vaziri Zadeh 2016).

State intervention is determining in these cases. Despite the evident inefficiency of the state in rent control and corruption prevention, the state is the most vigorous agent to manage an accountable monetary system of housing, distribution of welfare and public goods, and stability of housing market (for the benefit of the lower income). In the absence of an accountable developed system of funding that secures the provision of sufficient funds, the achievements of these programmes' objectives cannot be guaranteed, particularly about affordable housing.

(b) regeneration processes should adopt a long implementation period to avoid the destruction of local social structures:

Within a market-based framework, regeneration involves a process of reinvestment in an area after a long period of disinvestment. Rethinking the process of reinvestment is vitally important to deliver services to the low-income people and prevent dislocation. Porter and Shaw (2009: 247) argue that less fluctuations over longer periods of time are less damaging for the poor who are most vulnerable against massive and rapid reinvestment. Thus, a certain level of reinvestment would be desirable to benefit those who are least protected by wealth. The contemporary practices in Tehran reveal that all ambitious interventions have led to failure. Gradual renewal of the built and lived environment and adopting policies like 'rent control', would help neighbourhoods to improve their affordability according to changes. Policies that are based on long-term changes probably benefit the low-income more, avoid massive dislocation, and may better respect the rights of tenants.

(c) the participation of private sector is crucial, albeit with reservations:

Budgetary constraints are the main reason why these programmes do not achieve their objectives in Iran. The supply of long-term, reliable finance has been a major obstacle for (national and local) governments to implement regeneration programmes. Consequently, housing-led approach has become increasingly market-driven. If the regeneration processes adopt profit-oriented approaches, they will lead to gentrification (Leary and McCarthy 2013).

The integration of the private developers into programmes can facilitate access to funds required for the successful implementation of housing-led initiatives. However, the encouragement of private partnerships to invest in an area with the history disinvestment requires a variety of measures, including de-regulation (land-use readjustment), tax relief and low-interest loans. Paradoxically, the incentives for private funding probably contributes to a neglect of the main objectives of the regeneration programmes and result in social restructuring of the areas. The trade-off between the benefit making of the private sector and pursuing social objectives of these programmes is crucial.

(d) each programme is unique and through some innovative methods the programme has to involve the homegrown social context to encourage participation of people:

Social acceptability and the success of housing-led urban regeneration programmes can be improved through the involvement of the community and sufficient interactions with key actors. This has been ignored in almost entire history of regeneration in Tehran. The desire to breakup social exclusion and concentrations of deprivation (through housing-led initiatives) has to remain open to more community-based collaborative redevelopment policies that are closely allied to the aspirations, requirements and needs of the local people. People's participation in negotiations for regeneration results in a better achievement of a housing market that addresses both the affordability of the poor and the variety of social needs. However, the involvement of people in the countries with a long history of ignoring participation is very challenging. In this case, the role of local authorities is decisive. Instead of central government, local authorities (municipalities, city councils and particularly communities' councils) would manage and coordinate parties. The empowerment of councils and their engagement in the process of decision making indisputably improve the level of local involvement (more in: Moulaert et al. 2010).

(e) besides construction, the empowerment of deprived households through various methods has to be achieved and the programme has to be attuned to the welfare system objectives and social (local and national) policies:

In a wider perspective, we should remember that inequalities could not be tackled simply at neighbourhood level. There should be a closer connection between area-based approaches and national level main programmes (particularly the welfare social policies) (Allen 2006: 83). The remedy for urban deprived areas has to be sought not just in spatial reordering: the sources of inequality should also be tackled by rethinking local communities' access to social welfare (e.g. education, training and employment opportunities) in a more equitable city. The de-commodification of welfare services and resources is vital. Thus, the programmes must aim for developing a city equally accessible to all citizens (Arbaci and Rae 2013). In a larger context, housing-led regeneration programmes need to be mindful of moving social issues to nearby areas. It is of vital importance in housing-led regeneration programmes to retain the social structure of the area (avoiding some policies that led to densification, gentrification or engineering eviction).

13.5 Conclusion

The housing system in Iran has promoted an unhindered commodified market. It has left the land and property price to free market, inevitably leading to not only unmet housing needs of deprived households and massive urban intra- and extra-migration of social groups, but also a neglect of inner areas in historic zones leaving them entrapped in a cycle of decline.

Market failure has politically justified governmental interventions in the declining areas. Subsequently, the state-led and mostly ambitious policies for the intervention in the areas are implemented. Tackling the failure of commodified housing market during the last decade in redevelopment of declining areas, many controversial regeneration programmes were initiated to mitigate the vulnerability of built environment, to balance the housing market, to alleviate social exclusion, and to provide affordable housing for some deprived households.

There is evidence to show that Tehran's contemporary regeneration practices (even before the housing-led approach became mainstream) has always been associated with the housing system. However, there is controversy about the achievements of programmes, which contributed in reshaping social structure, dislocation of people and aggravate of exclusion. The chapter argues that its suggested considerations in urban regeneration policies in declining areas can improve the achievements of (affordable) housing-led programmes, increase affordability of housing, and most importantly, alleviate the socio-spatial segregation in the cities and preventing social exclusion.

References

Allen J (2006) Welfare regimes, welfare systems and housing in Southern Europe. Eur J Housing Policy 6(3):251–277
Arbaci S, Rae I (2013) Mixed-tenure neighbourhoods in London: policy myth or effective device to alleviate deprivation? Int J Urban Reg Res 37(2):451–479
Arbaci S, Tapada-Berteli T (2012) Social inequality and urban regeneration in Barcelona city centre: reconsidering success. Eur Urban Reg Stud 19(3):287–311
Azizi MM (1998) Evaluation of urban land supply policy in Iran. Int J Urban Reg Res 22(1):94–105
Cameron S (2003) Gentrification, housing redifferentiation and urban regeneration: "Going for Growth" in Newcastle upon Tyne. Urban Stud 40(12):2367–2382
Cameron S (2006) From low demand to rising aspirations: housing market renewal within regional and neighbourhood regeneration policy. Housing Stud 21(1):3–16
Cameron S, Doling J (1994) Housing neighbourhoods and urban regeneration. Urban Stud 31(7):1211–1223
Drakakis-Smith D (1981) Housing and the urban development process. Croom Helm. Duncan, London
Ennis F(2016) Infrastructure provision and the negotiating process: urban and regional planning and development series (F Ennis, ed). Routledge, Chapman & Hall, Incorporated
Etemad G et al (2012) Evaluation of Navvab Project and its consequences (Arzyabi-e-Tarh-e-Navvab va Payamadhay-e-Aan). Maani Publisher, Tehran
Gutiérrez Romero R (2009) Estimating the impact of England's area-based intervention 'New Deal for Communities' on employment. Reg Sci Urban Econ 39(3):323–331
Healey P, Davoudi S, Tavsanoglu S (1992) Rebuilding the city: property-led urban regeneration. Spon, London
Imrie R, Lees L, Raco M (eds) (2009) Regenerating London: governance, sustainability and community in a global city. Routledge, London and New York
Jones P, Evans J (2008) Urban regeneration in UK. Sage, London
Kamrava (2010) Evaluation of the last 50 years regeneration practices (barresiye 50 sal tajrobeye nosazi)
Keivani R, Werna E (2001) Refocusing the housing debate in developing countries from a pluralist perspective. Habitat Int 25(2):191–208
Lawless P (2006) Area-based urban interventions: rationale and outcomes: the new deal for communities programme in England. Urban Stud 43(11):1991–2011
Lawless P (2012) Can area-based regeneration programmes ever work? Evidence from England's new deal for communities programme. Policy Stud 33(4):313–328
Lawless P, Foden M, Wilson I, Beatty C (2009) Understanding area-based regeneration: the new deal for communities programme in England. Urban Stud 47(2):257–275
Leary ME, McCarthy J (2013) The Routledge companion to urban regeneration (ME Leary and J McCarthy, eds). Routledge, Abingdon and New York
Lupton R, Tunstall R (2008) Neighbourhood regeneration through mixed communities: a 'social justice dilemma'? J Educ Policy 23(2):105–117

MacInnes T, Aldridge H, Bushe S, Kenway P, Tinson A (2013) Monitoring poverty and social exclusion 2013: summary report. Joseph Rowntree Foundation. Retrieved from

Madanipour A (2006) Urban planning and development in Tehran. Cities 23(6):433–438

McCarthy J, Pollock SA (1997) Urban regeneration in Glasgow and Dundee: a comparative evaluation. Land Use Policy 14(2):137–149

Mcgregor A, McConnachie M (1995) Social exclusion, urban regeneration and. Urban Stud 32 (10):1587–1600

Moulaert F, Martinelli F, Swyngedouw E, González S (eds) (2010) Can neighbourhoods save the city? Community development and social innovation. Routledge, London and New York

Parry JM, Laburn-Peart K, Orford J, Dalton S (2004) Mechanisms by which area-based regeneration programmes might impact on community health: a case study of the new deal for communities initiative. Public Health 118(7):497–505

Porter L, Shaw K (eds) (2009) Whose urban renaissance? an international comparison of urban regeneration strategies. Routledge, London

Pugh C (1997) Poverty and progress? Reflections on housing and urban policies in developing countries, 1976–96. Urban Stud 34(10):1547–1595

Pugh C (2001) The theory and practice of housing sector development for developing countries. Housing Stud 16(4):399–423

Rezayee R (2009) Evaluation of the regeneration programme in Khobbakht Neighbourhood (arzyabi-e-tarh-e-nosaziy-e-mahaley-e-Khobbakht)

Roberts P, Sykes H, Granger R (eds) (2016) Urban regeneration (2nd ed)

Robertson DS (1998) Pulling in opposite directions: the failure of post war planning to regenerate Glasgow. Plann Perspect 13(March):53–67

Sazman-e-Nosazi-e-Shahr-e-Tehran (2015) Documentation of flagship projects in Tehran (Mostanad negari projehaye shakhese shahre Tehran). Tehran

Scanlon K, Whitehead C (2008) Social housing in Europe II: a review of policies (K Scanlon, C Whitehead, eds). London School of Economics and Political Science, London

Sepehrdoust H (2012) The impact of migrant labor force on housing construction of Iran. J Housing Built Environ 28(1):67–78

Smith GR (1999) Area-based initiatives: the rationale and options for area targeting. LSE STICERD Research Paper No. CASE025. London

Tallon A (2009) Urban regeneration in the UK. Routledge, London

Tunstall R, Lupton R (2003) Is targeting deprived areas an effective means to reach poor people? An assessment of one rationale for area—based funding programmes. London

Vaughan L, Arbaci S (2007) The challenges of understanding urban segregation. Built Environ 37 (2):128–138

Vaziri Zadeh A (2016) The evaluation of state involvement in large-scale property-led regeneration projects in Iran. In: Arefian FF, Moeini SHI (eds) Urban change in Iran. Springer International Publishing, Switzerland, pp 215–229

World Bank (1994) World development report 1994. Infrastructure for development. World Development (vol 26). Oxford University Press, Oxford and New York

Chapter 14
Integrative Conservation of Tehran's Oldest Qanat by Employing Historic Urban Landscape Approach

Narjes Zivdar and Ameneh Karimian

Abstract *Qanats* have played a vital role in underground water extraction since ancient times in *Iran*. They also have a special niche in the cultural, social, economic, political and physical landscapes of the country. There are 50,000 qanats in *Iran*, more than 200 of which, with a total length of 2,000 km, are in *Tehran*, the capital city. Over the past decades, Tehran's urban core is exposed to development pressures which adversely affected the urban green infrastructures and specifically the qanats network. *Mehrgerd* is the oldest active, but also endangered qanat in Tehran, which continues to drain groundwater without any maintenance for over 60 years. However, it is nearly abandoned and needs to be conserved. In recent experiences on revitalisation of historic qanats in Iran, the focus has mainly been on physical restoration, with community engagement and sociocultural aspects mostly overlooked. This research aims at developing a process to enhance conservation of Mehrgerd qanat by rereading and rethinking its position employing Historic Urban Landscape (HUL) approach as set out in the 2011 UNESCO's recommendation on this subject. To do so, the method is explained, and then, the main layers are defined and analysed based on the Heritage Impact Assessment' (HIA) method. This mapping exercise finally leads to overlay weighted value layers for Mehrgerd's historic urban landscape. The paper concludes that this assessment framework can be used for determining the priorities and the level of conservation in Mehrgerd green infrastructure in integration with its nearby historic urban landscape in order to preserve this exceptional urban heritage more efficiently.

Keywords Tehran's qanat · Mehrgerd · Integrative conservation · Historic urban landscape · Heritage impact assessments

---

N. Zivdar (✉)
United Nations Representation, Tehran, Iran
e-mail: narjeszivdar@gmail.com

A. Karimian
Shahid Beheshti University, Tehran, Iran
e-mail: ameneh.karimian@gmail.com

© Springer Nature Switzerland AG 2020
F. F. Arefian and S. H. I. Moeini (eds.), *Urban Heritage Along the Silk Roads*,
The Urban Book Series, https://doi.org/10.1007/978-3-030-22762-3_14

14.1 Introduction

Throughout the arid regions of Iran, human settlements were supported by the ancient qanat system of tapping alluvial aquifers at the heads of valleys and conducting the water along underground tunnels by gravity, often over many kilometres (Fig. 14.1). Appearing as craters from above, shaft wells provided access and ventilation to the tunnels, following the route of the qanat from water source to settlements (Fig. 14.2) (ICHHTO 2015). These exceptional green infrastructures depend not only on sophisticated engineering knowledge, but also on a refined communal management system.

A city of 13 million in the desert and like most cities in Iran, Tehran is located at the foot of a mountain in order to source water. Tehran's historic core is geo-strategically positioned in the axis of Towchal mountain summit. The five main water valleys that absorb the mountain's snowfall are the roots of the capital's ancient qanat network. There are over 200 qanats in Tehran with a total length of 2000 km and 80,000 wells which drain 10,000 litres of water per second (Kamalvand et al. 2016) and providing water for the green spaces nearby; although some of them are currently non-operational (ICHHTO 2015). According to Henri Goblot, the time of Qajar can be considered as the heyday of qanats (Goblot 1979). Agha Mohammad Khan, the founder of the Qajar dynasty chose Tehran as his capital city: a city with no access to a reliable stream of surface water and totally reliant on groundwater. The rich supply of groundwater and suitable geological-topographical conditions of Tehran allowed this city to house many

Fig. 14.1 A view of qanat in arid regions of Iran, Sistan and Balouchestan (Authors)

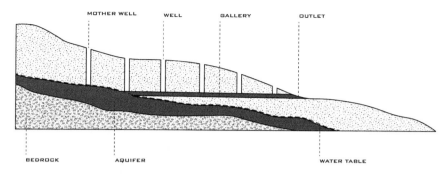

Fig. 14.2 Schematic section of Qanat structure (Authors)

Fig. 14.3 Qanat network and the green infrastructure of Tehran capital city (Authors)

qanats and gave birth to the city (Semsar Yazdi and Askarzadeh 2007, p. 318) (Fig. 14.3).

According to orally quoted information verified by experts, Mehrgerd is the oldest operational qanat in Tehran. Located at the heart of the historic city (Fig. 14.4). During its 700-year history (Hensel and Gharleghi 2012), it has provided water and still supplies water for irrigating some urban public spaces. It still drains 200 litres of groundwater per second by gravity and without any maintenance for the past 60 years (Kamalvand et al. 2016). Due to drastic, uncontrolled urban changes such as the development of Tehran's subway and other modern infrastructures, some irreversible changes have occurred in the qanat's physical structure. Moreover, as a result of changes in people's lifestyle and the

Fig. 14.4 Mehrgerd qanat location in Tehran historic urban context (Authors based on the historic maps of Tehran)

emergence of modern urban water supply systems and no use of qanat for drinking water supply, its communal management system and consequently sociocultural importance has vanished over the time. All these changes adversely affected this qanat and endangered its exceptional urban green infrastructure.

14.2 Objectives and Method

The main objective of this research is developing a process to define a conservation framework for safeguarding Mehrgerd qanat based on 'UNESCO's 2011 Recommendation on Historic Urban Landscapes' (HUL). For this purpose, definitions, principles and tools of HUL are first explained. The main information and data related to Mehrgerd qanat and its historic urban context have been collected based on a qualitative approach associated with analytical methods and in-depth surveys and interviews with current stakeholders as well as taking note of existing reports and researches. To this end, the information available is organised in layers in line with tools recommended in HUL document. Then, the main layers including historical, cultural, social, economic, infrastructural and environmental have been assessed based on grading scales as set out in '2011 ICOMOS guidance on Heritage Impact Assessments' (HIA) for classification and assessing the impacts and the significance of heritage assets and attributes in the world heritage context of Mehrgerd. This analytic mapping exercise finally leads to establishing value layers for Mehrgerd qanat historic urban landscape. Value layers are then developed by overlaying weighted layers in a Geographic Information System (GIS). The resulting assessment framework leads to integrative conservation of Tehran's oldest qanat.

14.3 Area of Work

Mehrgerd qanat path (Fig. 14.5) starts from its mother wells, one in Behjat Abad urban park and the other in Zartosht Street, and continues towards the historic centre of Tehran (Fig. 14.6). Before the modification, the qanat's emerging point was near Golestan Palace and Garden with its other branches distributed in Tehran Historic bazaar. Golestan Palace is the only World Heritage site in the city and the Bazaar is the socio-economic centre of Tehran capital city. Therefore, Mehrgerd qanat's location is very strategic in terms of environmental, historical, social, cultural and economic significance. On top of that, the qanat is not only an isolated underground system with the monofunctional use of bringing water to the surface

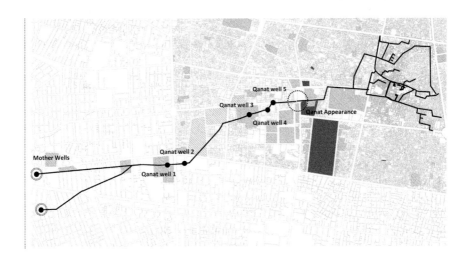

Fig. 14.5 Schematic map of Mehrgerd Qanat (Authors)

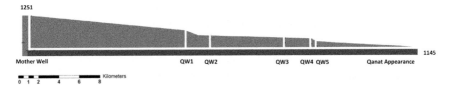

Fig. 14.6 Schematic section of Mehrgerd Qanat (Authors based on Hydrocity project)

through a linear path, but also an infrastructure related to different parts and layers of urban context. So, in this research the field study and data gathering have not been restricted to the qanat path itself and the mapping exercise of the layers are rather conducted in the qanat urban context in order to achieve a comprehensive understanding of this exceptional green infrastructure.

14.4 Historic Urban Landscape (HUL)

Over the past decades, the definition of heritage management has been evolving from an object-based approach towards a more all-inclusive approach (Veldpaus and Roders 2013). A landscape approach, such as the Historic Urban Landscape approach, is expected to be a future path in heritage management, as well as, a key indicator for sustainable urban development (Veldpaus and Roders 2013). The recent UNESCO (2011) recommendation on the Historic Urban Landscape (HUL) provides guidance on such a landscape-based approach at international level. HUL is UNESCO's approach to managing historic urban landscapes that is holistic by integrating the goals of environmental conservation and those of social, cultural and economic development with an emphasis on a broader appreciation of socio-economic processes in the conservation of urban environment values rather than the limited focus on architectural monuments or ensembles (Bandarin and Oers 2012).

The concept of HUL as quoted in the UNESCO Recommendation is:

> The urban area understood as the result of an historic layering of cultural and natural values and attributes, extending beyond the notion of 'historic centre' or 'ensemble' to include the broader urban context and its geographical setting.

The historic urban landscape approach aims at preserving the quality of the human environment, enhancing the productive and sustainable use of urban spaces while recognising their dynamic character, and promoting social and functional diversity. It is rooted in a balanced and sustainable relationship between the urban and natural environments, between the needs of present and future generations and the legacy from the past. The historic urban landscape approach considers cultural diversity and creativity as key assets for human, social and economic development and provides tools to manage both environmental and socio-economic transformations and to ensure that contemporary interventions in the urban environment are harmoniously integrated with natural and cultural heritages in a historic setting and take into account regional contexts (World Heritage Centre 2011).

The historic urban landscape approach learns from the traditions and perceptions of local communities while respecting the values of the national and international communities (World Heritage Centre 2011).

As suggested in the 2011 Historic Urban Landscape Recommendation, 'Urban heritage, including its tangible and intangible components...' must be understood as embodying:

the site's topography, geomorphology, hydrology and natural features; its built environment, both historic and contemporary; its infrastructures above and below ground; its open spaces and gardens, its land use patterns and spatial organisation; perceptions and visual relationships; as well as all other elements of the urban structure. It also includes social and cultural practices and values, economic processes', which implies the tangible dimensions of heritage assets, 'and the intangible dimensions of heritage as related to diversity and identity (Bandarin and Oers 2015).

According to the above-mentioned definition, the concept of HUL is not something new, what is new is the changed perception about the immense potential of this concept, the possibility of treating urban areas not as static objects of admiration but as living spaces for resilient communities and sustainable urban environments (Martini 2013).

14.5 Heritage Impact Assessments (HIA)

Impact assessment has been a tool used by other sectors, in particular the environment sector, and gradually being introduced into the heritage sector (Fig. 14.7), supporting assessments of both natural and cultural heritages in the urban areas. Heritage Impact Assessment is a process of identifying, predicting, evaluating and communicating the probable effects of a current or proposed development policy or action on the heritage values (including outstanding universal value in the case of

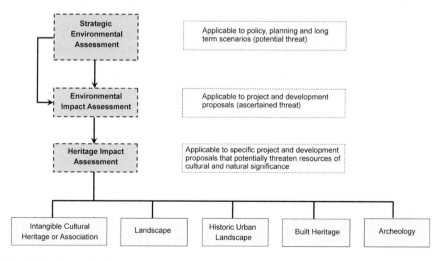

Fig. 14.7 The main fields indicated by Heritage Impact Assessments, HIA (Authors based on ICOMOS 2011; Roders and Oers 2012)

World Heritage Properties), cultural life, institutions and resources of communities, then integrating the findings and conclusions into the planning and decision making process, with a view to mitigating adverse impacts and enhancing positive outcomes (International Association for Impact Assessment 2002).

In recent years the UNESCO World Heritage Committee has addressed considerable numbers of State of Conservation Reports related to threats to World Heritage properties from various forms of urban development. These developments include contextual or insensitive developments, renewals, demolitions and new infrastructure typologies. Many of these projects have had the potential to impact adversely on different attributes in the world heritage sites (ICOMOS 2011) and specifically the historic and environmental infrastructures in urban areas which have been occurred in the Mehrgerd historic urban context as well.

In order to satisfactorily evaluate these potential threats, there is a need to be specific about the impacts of proposed changes in these areas (ICOMOS 2011). While heritage impact assessment exists in many countries like Iran, these seem less reliably used in the World Heritage context. There are also limited formal tools for identifying receptors and assessing impact and few examples of excellence for Heritage Impact Assessment (HIA) undertaken for cultural WH properties. In order to define an integrative conservation plan, ICOMOS released 'Guidance on Heritage Impact Assessments for Cultural World Heritage Properties, HIA' proposing an efficient framework and tools (process, classification, evaluation methods) in assessing the impacts of the Outstanding Universal Value, OUV in 2011. Through the procedure of HIA, the significance of the effect of change, i.e. the overall impact on an attribute is a function of the importance of the attribute and the scale of change. Based on this plan, the monitoring system will have clear indicators to implementation (Yen and Cheng 2015) in different contexts including archaeological or built heritage assets or associated with natural and cultural heritage as well as landscapes and historic urban landscapes (Fig. 14.7).

HIAs for WH properties will need to consider their international heritage value and also other local or national values, and priorities or recommendations set out in national research agendas. They may also need to consider other international values reflected in the property. The value of the asset may be defined using the following grading scale (Table 14.1).

Table 14.1 HIA grading scale (Authors based on ICOMOS 2011)

Grading scale	Value
Very high	5
High	4
Medium	3
Low	2
Negligible	1
Unknown potential	0

Fig. 14.8 Mapping historic and cultural value layer (Authors), A Golsetan Palace

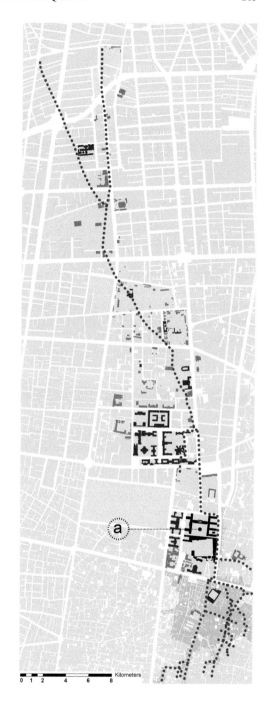

14.6 Mehrgerd Historic Layering of Values

In line with the layering concept and the tools of HUL recommendation, the information and data about the Mehrgerd qanat landscape is organised in five main layers including historical, cultural, social, economic, pathways, infrastructural and environmental layers, and then assessed based on the grading scales as set out in '2011 ICOMOS guidance on Heritage Impact Assessments' (HIA) (Table 14.1). Followings are Mehrgerd's HUL value layers.

14.6.1 Historic and Cultural Value Layer

As mentioned before, Golsetan Palace (Fig. 14.8a) is the only World Heritage property in Tehran, which has the highest grade regarding historic and cultural value layer. The National Garden complex with its museums and buildings, and some other historic buildings along Mehrgerd path—which have mostly turned into museums—are also inscribed as National Cultural Heritage properties. The bazaar and some other iconic buildings with exceptional architecture and history values make the medium grade (Table 14.2).

14.6.2 Socio-economic Value Layer

Tehran's grand bazaar roots back to sixteenth century and have played a vital role in socio-economic and political history of Iran. Before making the changes in

Table 14.2 HIA for historical and cultural value (Authors based on ICOMOS 2011)

Built heritage or HUL, historic landscape, intangible cultural heritage	Grading	#
Area wholly inscribed on the buffer zone of World Heritage property. (Golestan Palace buffer zone) or associated with Intangible cultural heritage activities as evidenced by the national register	Very high	5
Nationally inscribed buildings or areas containing important buildings (National Cultural Heritage organisation list or buildings that have exceptional qualities or historical associations not adequately reflected in the listing grade)	High	4
Historic (unlisted (buildings with exceptional qualities or historical and cultural associations)	Medium	3
Historic landscapes or Buildings with importance to local interest groups or associated with Intangible cultural heritage activities of local significance	Low	2
Buildings or urban landscapes of no architectural, cultural or historical merit	Negligible	1
Areas with some hidden (inaccessible) potential	Unknown	0

Fig. 14.9 Mapping socio-economic value layer (Authors), B Tehran grand bazaar

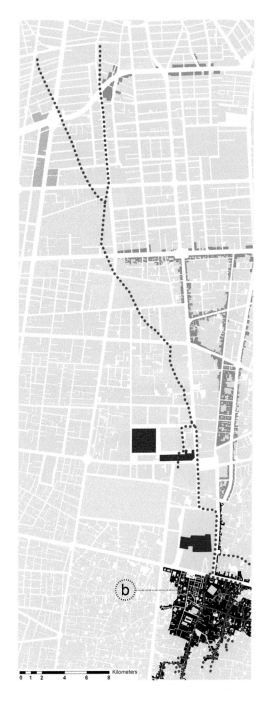

Table 14.3 HIA for socio-economic value (Authors based on ICOMOS 2011)

Built heritage or HUL, historic landscape, intangible cultural heritage	Grading	#
Area with exceptional socio-economic value and resources of national importance. (Tehran historic Bazaar)	Very high	5
Area with high socio-economic value and resources of national importance	High	4
Area with Medium socio-economic value and resources of regional importance	Medium	3
Area with Low socio-economic value and resources of local importance	Low	2
Area with no significant socio-economic resources	Negligible	1
The socio-economic importance of the area has not been ascertained	Unknown	0

Mehrgerd qanat path due to vast modern construction, it used to run in Bazaar alleys and provide potable water. It still lives in collective memory of Bazaar's old people. Consequently, the grand bazaar (Fig. 14.9b) is marked as 'very high' in socio-economic value layer. Other than that, some the country's major museums including Golestan and the National Museum play a significant role in the social and economic life of the city and are graded as 'high'. Other markets which are more regional like Behjat Abad, Laleh Zar and Karim Khan are rated medium in this layer (Table 14.3).

14.6.3 Environmental Value Layer

Park-e Shahr (Fig. 14.10c) as one of the city's most important urban parks, the National Garden (Fig. 14.10d) and *Golestan Palace* with their striking open spaces along with *Sabzeh Meidan* (a small open space in the Grand Bazaar) and *Abgineh* Museum have the very high level grading in environmental value layer. Other urban parks, green spaces and open spaces in the area are rated according to their importance and roles as mentioned below (Table 14.4).

14.6.4 Pathway Value Layer

The passing of the qanat from underneath *Enqelab*, *Jomhouri* and *Imam Khomeini* and the nearby *Valiasr* Avenues is another point to highlight the strategical buffer zone of the Mehrgerd pathway (Fig. 14.11 and Table 14.5).

Fig. 14.10 Mapping environmental value layer (Authors), D Park Shahr, C National Garden

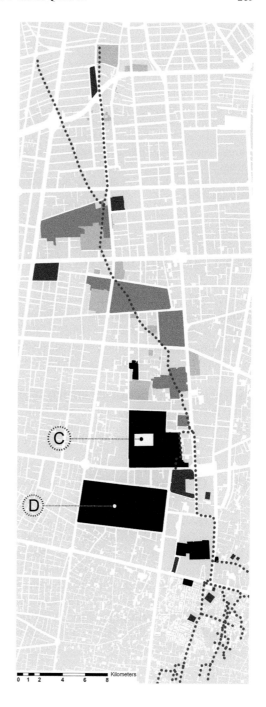

Table 14.4 HIA for environmental value (Authors based on ICOMOS 2011)

Built heritage or HUL, historic landscape, intangible cultural heritage	Grading	#
Historic Garden, open and green space or landscape of international value with exceptional values such as coherence and time depth. (Persian garden)	Very high	5
Nationally designated historic landscape or open space associated with globally important Intangible cultural heritage activities	High	4
Designated special historic green space or public landscape of regional value, associated with Intangible Cultural heritage activities	Medium	3
Historic open or green space or landscape with importance to local interest groups	Low	2
Public spaces or landscapes with little or no significant natural resources or historical interest	Negligible	1
N/a	Unknown	0

14.6.5 Infrastructure Value Layer

Before 1993, Mehrgerd's water used to run all the way from its mother wells to the Grand Bazaar alleys. After the 1993 locating of Imam Khomeini subway interchange (Fig. 14.12e) (the main station of the city) on Mehrgerd pathway, the qanat channel was diverted, sending water to waste from that point (Fig. 14.12f): a sort of water that based on recent research has a very high quality and can be used for irrigation with no need for filtration (Kamalvand et al. 2016). Mehrgerd qanat as the focal point of this research has the very high ranking in infrastructure value layer, with intersecting subway lines rated high accordingly (Figs. 14.13 and 14.14, Table 14.6).

14.7 Conclusions

After assessing the heritage impact of each layer in Mehrgerd qanat historic urban landscape, the value layers are developed by overlaying weighted layers in a Geographic Information System (GIS). As a result, the most significant areas along the qanat rout have been recognised (Fig. 14.15) and accordingly it shows the most vulnerable zones which must be considered as priorities in qanat revival and conservation. In a way of an example, according to the overlaid value map (Fig. 14.15g) the former appearance of Mehrgerd qanat is in fact one of the most important spots of this green urban infrastructure which is in front of Golestan

Fig. 14.11 Mapping pathway value layer (Authors)

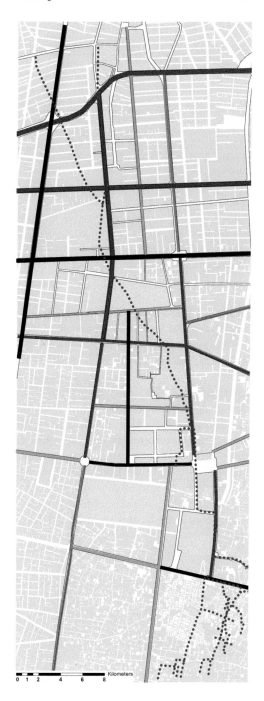

Table 14.5 HIA for pathway value

Built heritage or HUL, historic landscape, intangible cultural heritage	Grading	#
Pathway with exceptional coherence, time depth or association with Intangible cultural heritage activities as evidenced by the national register or of global significance. (Pathways associated with activities related to ritual dramatic art of Ta'zīye in Ashoura annual traditional event)	Very high	5
Pathway of national importance includes in conservation areas or pathways that can be shown to have exceptional qualities in their fabric or historical associations not adequately reflected in the listing grade or the one associated with Intangible cultural heritage activities of national significance	High	4
Pathway of regional value associated with Intangible Cultural heritage activities as evidenced by local registers	Medium	3
Pathway of local value with poor survival of historic and contextual associations	Low	2
Pathway with poor survival of contextual associations	Negligible	1
The importance of the pathway has not been ascertained	Unknown	0

Palace and Garden World Heritage Site and despite its high significance is highly endangered and should be preserved.

This research indicates that Mehrgerd qanat as the oldest green infrastructure in Tehran capital city cannot be revitalised properly unless other layers, aspects and attributes of urban heritage, and especially those of social, cultural and economic nature are taken into account coherently. In the light of the dynamic process of overlaying HUL layers, the urban environment has been considered as a living organism: the result of a historic layering of values, which can adapt itself to the necessities of modern lifestyle is seen in a developed perspective which should be based on the balance between conservation and development, the past and the future of historic urban landscapes. Finally, this research aims at developing this dynamic process towards a general framework for integrative conservation of urban green infrastructures and specifically qanats network in Tehran, taking note of the results concluded by this research and future applications of this methodology as well.

Fig. 14.12 Mapping infrastructural value layer (Authors), E Imam Khomeini subway station. F Qanat water drainage point

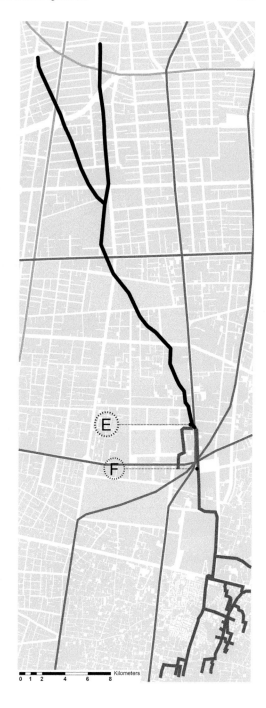

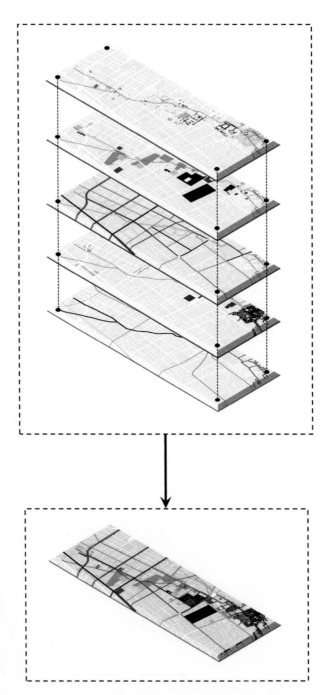

Fig. 14.13 The dynamic process of overlaying weighted value layers for Mehrgerd's HUL (Authors)

14 Integrative Conservation of Tehran's Oldest Qanat … 225

Fig. 14.14 Mehrgerd Qanat surrounding heritage assets (photos taken by the Authors). **a** Alborz College pool. One of the few places that still utilises the water of Mehrgerd Qanat. **b** Alborz College, an iconic building along the path of Mehrgerd Qanat. **c** Behjat Abad Bazaar, an old bazaar along the path of Mehrgerd Qanat. **d** Yarjani St., a paved street downtown in the strategic area of National Garden, close to Mehrgerd Qanat path

Table 14.6 HIA for infrastructure value

Built heritage or HUL, historic landscape, intangible cultural heritage	Grading	#
Exceptional infrastructural assets of international importance inscribed as WH property, associated with particular innovations of global significance. (Persian Qanat network in Tehran historic core: Mehrgerd Qanat)	Very high	5
Infrastructural element Associated with particular technical developments of national significance. (Tehran subway)	High	4
Other infrastructures that can be shown to have exceptional qualities in their fabric or historical associations not adequately reflected in the listing grade	Medium	3
Infrastructural element of regional significance	Low	2
Infrastructural element of local significance	Negligible	1
N/a	Unknown	0

Fig. 14.15 Overlaying weighted value layers (Authors), G Mehrgerd qanat appearance in front of Golestan Palace World Heritage Site

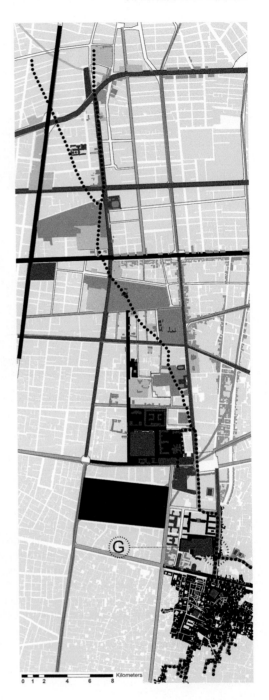

References

Bandarin F, van Oers R (2012) The historic urban landscape: preserving heritage in an urban century. Wiley-Blackwell, Oxford, UK

Bandarin F, van Oers R (eds) (2015) Reconnecting the city: the historic urban landscape approach. Wiley-Blackwell, Chichester

Goblot H (1979) Les qanats: Une technique d'acquisition de l'eau. Ecole Hautes Etudes Sciences Sociales, Paris

Hensel M, Gharleghi M (2012) Iran: past, present and future. Wiley, London

ICHHTO (2015) The Persian Qanat: for inscription on the world heritage list (Executive Summary), UNESCO World Heritage Centre. http://whc.unesco.org/uploads/nominations/1506.pdf

ICOMOS (2011) Guidance on heritage impact assessments for cultural world heritage properties. Paris, International Council on Monuments and Sites (ICOMOS)

International Association for Impact Assessment (2002) EIA training resource manual: Topic 13-Social Impact Assessment. http://www.iaia.org/pdf/UNEP/Manualcontents/EIA_E_top13_body.pdf

Kamalvand S, Karimian A, Zivdar N, Eskandari S, Jamali F, Hemmati M, Meschi M (2016) Tehran garden festival: mapping mehrged (unpublished). Tehran: an applied research project by HydroCity at École Spéciale d'Architecture in Paris

Martini V (2013) The conservation of historic urban landscapes: an approach. Unpublished doctoral dissertation, University of Nova Gorica

Roders AP, van Oers R (2012) Guidance on heritage impact assessments: learning from its application on world heritage site management. J Cultural Heritage Manag Sustain Dev 2(2):104–114

Semsar Yazdi A, Askarzadeh S (2007) A historical review on the Qanats and historic hydraulic structures of Iran since the First Millennium B.C., International History Seminar on Irrigation and Drainage, Tehran May 2–5, 2007

Veldpaus L, Roders AP (2013) Historic urban landscape: an assessment framework. In: Lecture presented at 33rd annual meeting of the international association for impact assessment, Calgary Stampede BMO Centre, Calgary, Alberta, Canada, 13–16 May 2013

World Heritage Centre (2011) UNESCO's recommendation on the historic urban landscape. United Nations Educational, Scientific and Cultural Organisation (UNESCO). http://whc.unesco.org/en/activities/638

Yen YN, Cheng CF (2015) Conservation plan based on the concept of integrity, In: Yen YN, Weng KH & Cheng HM (eds) ISPRS Annals of the Photogrammetry. The Proceeding of 25th International CIPA Symposium, Taipei, Taiwan, 31 August–4 September 2015

Chapter 15
Assessing the Pedestrian Network Conditions in Two Cities: The Cases of Qazvin and Porto

Mona Jabbari, Fernando Pereira da Fonseca and Rui António Rodrigues Ramos

Abstract The quality of life in cities depends on the existence of suitable conditions to walk. The aim of this chapter is to assess the conditions provided to pedestrians in two cities with different urban morphologies: Qazvin (Iran) and Porto (Portugal). The assessment was performed through a model that combines multi-criteria analysis with street network connectivity to evaluate the pedestrian conditions. The multi-criteria analysis was carried out by using four criteria and nine sub-criteria that mostly influence walkability and by involving a group of experts from Qazvin and Porto. Street network connectivity was assessed by Space Syntax. Results showed that Qazvin provides better conditions and a network of pedestrian streets more connected than Porto. The model can be a useful tool for planning more walkable and sustainable cities in urban areas.

Keywords Walkability · Pedestrian network · Multi-criteria analysis · Street network connectivity analysis

15.1 Introduction

Walking is one of the least expensive and most broadly accessible modes of transportation. Walkability provides the opportunity for large numbers of people to walk together and experience the route or site (Dallman et al. 2013; Fernando et al. 2010; Jamei and Rajagopalan 2017; Socharoentum and Karimi 2016). This can be a powerful way to build a sense of community and to strengthen social networks. Pedestrian environments (sidewalks, paths and hallways) are a major portion of the

---

M. Jabbari (✉) · F. P. da Fonseca · R. A. R. Ramos
University of Minho, Braga, Portugal
e-mail: mona.jabbari@civil.uminho.pt

F. P. da Fonseca
e-mail: ffonseca@civil.uminho.pt

R. A. R. Ramos
e-mail: rui.ramos@civil.uminho.pt

public realm. Many beneficial activities (socialising, waiting, shopping and eating) occur in pedestrian environments (Buccolieri et al. 2015; Ferreira et al. 2016; Green and Klein 2011). Recent research shows that commercial activity, tourism, information technology and leisure economy are fostering economic growth (Evans 2001; Farrell 2000; Gospodini 2006). Walking supports this development and becomes conducive to create economic value and social vibrancy.

This chapter describes the result of the Pedestrian Network Assessment (PNA) developed for assessing the pedestrian conditions provided by the cities of Qazvin and Porto. The model was underpinned in a multi-criteria analysis and in street connectivity analysis. Through multi-criteria analysis, different weights were assigned to four criteria and nine sub-criteria with impact in walking. The weights were assigned considering the results of a survey where respondents evaluated the importance of such criterion in a scale ranging from 0 (minimal weight) to 1 (maximum weight). Street network connectivity (SNC) was evaluated through Space Syntax analysis. The scores obtained were introduced in a GIS software (ArcGIS) by using a weighted linear combination (WLC). The PNA was applied in two cities with different urban morphologies: Qazvin (Iran) and Porto (Portugal). The main structure of Qazvin is polycentric and space was created due to activities and the neighbourhood. Also, the city centre of Porto has a strong spatial articulation which is known as monocentric. These two cities were selected to show the impact that different urban structures may have on pedestrian conditions. It is hypothesised that 'the more a space is integrated, the greater the chances that it will be cohesive pedestrian network' (Jeong and Banyn 2016; Li et al. 2016). In other words, the chapter will study the impact of street configuration on the walking conditions provided to pedestrians. The results obtained with the implementation of the PNA model in both cities are compared by using a quadrant chart method.

The chapter starts with a literature review in Sect. 15.2, followed by descriptions of the two case studies in Sect. 15.3. The methodology is presented in Sect. 15.4, the results are described in Sects. 15.5 and 15.6 and the conclusions are presented in Sect. 15.7.

15.2 Literature Review

There has been a dramatic rise in the literature on factors affecting walking in the last two decades. Researchers have pointed out several criteria affecting pedestrian movement (Badland et al. 2013; Cervero and Kockelman 1997; Havard and Willis 2012). One of these criteria is to do with land use, which determines the trip destination (Bahrainy and Khosravi 2013; Lamíquiz and López-Domínguez 2015; Lerman and Omer 2016). In addition, the residential population density is an important criterion that shows the amount of movement and depends on daily demand (Grecu and Morar 2013; Lerman and Omer 2016). Many pedestrian studies found in literature are related with behavioural aspects associated with physical environment including human scale, visual dimension and the route slope

(Bahrainy and Khosravi 2013; Forsyth et al. 2009; Gilderbloom et al. 2015; Lamíquiz and López-Domínguez 2015; Marquet and Miralles-Guasch 2016; Mehta 2008; Nasir et al. 2014; Peiravian et al. 2014). Walking to public transport and having access to its information have been encouraged as an active living strategy in many countries (Cubukcu et al. 2015) . However, accessibility is strongly linked with urban functions and physical environment (Gilderbloom et al. 2015; Lamíquiz and López-Domínguez 2015). The natural environment forms another criterion recognised for its effect on walkability thereby improving pedestrian movement (Lundberg and Weber 2014; Panagopoulos et al. 2016). Comfortable microclimatic conditions, including temperature, green space, sunlight shade and wind, are also important in supporting outdoor activities and walking (Koh and Wong 2013; Mehta 2008; Zadeh 1978). Finally, considerable research has focused on the relationship between walking and the above criteria, which contribute to an overall perception of walkability. Actually, these attributes are strongly associated with the decision to, and with the satisfaction of walking (Jabbari et al. 2017; Martinelli et al. 2015; Moura et al. 2017; Socharoentum and Karimi 2016).

On the other hand, the connectivity between pedestrian streets is another important attribute. Over the last years, street network connectivity analysis has contributed to a greater understanding of the spatial configuration of street networks and to the location of economic activities (Hillier and Hanson 1998; Hillier et al. 1993). Space Syntax is a tool often used for measuring the number of street intersections per line. A connection graph is defined depending on how each line connects to its surrounding lines (Jiang and Liu 2009; Penn et al. 1998).

Cities are clearly more complex than regularly structured systems; their geometry is variable and irregular. In turn, urban morphology is the primary generator of pedestrian movement patterns (Hillier et al. 1993). Peponis et al. (1989) presented some findings about morphology of Greek towns and their patterns of pedestrian movement. Some researchers have relied on general Graph theory to simulate the pedestrian network, which is detected in the conceptualisation of the patterns of urban morphology (Hillier and Iida 2005; Jayasinghe et al. 2016; Lerman et al. 2014; Li et al. 2016; McCahill and Garrick 2008; Önder and Gigi 2010). In this sense, urban morphology is another parameter for assessing the effects of urban form on walkability.

15.3 Case Studies: Porto and Qazvin

This study was developed comparing the Portuguese city of Porto and the Iranian city of Qazvin. These cities have different urban structures. The city centre of Porto has a strong spatial articulation with its urban morphology based on open space and urban tissues formed around it. In Porto, the urban structure has transformed during history, and finally organised around the Aliados Avenue. The main structure of Qazvin is different from Porto. The city is one of Silk Roads cities with each neighbourhood having boundaries defined by main streets. The urban structure of

Qazvin is comprised of several centres presenting a multinuclear morphology. Each neighbourhood centre was the local urban public space and consisted of a mosque, a water reservoir, a public bath and a flexible open space which was used for religious rituals as well as for daily social interactions (Pourjafar et al. 2014; Tavassoli 2016). Figure 15.1 shows the urban structures and the historical growth processes in of Porto and Qazvin.

The area selected in Porto for implementing the PNA comprises the historic centre of Porto and the neighbouring quarters. The area selected represents 7% of the city area, has about 2.6 km^2, and is defined by the streets of Santa Cantarina, Bovista, Casa de Música Square, Júlio Dinis, Palácio de Cristal, and by the Douro river. The selected are includes open spaces, sloped terrains, riverfront, green areas and historical fabric. These spaces provide a variety of places, turning the places into a varied and multifunctional space for pedestrians. The area for implementing the PNA in Qazvin is limited by Bou Ali Sina Street, Molavi Street, Shahid Ansari Street, Hokm Abad Garden and Boulevard. The area of the framework corresponds to 15% of the Qazvin city, which is about 2.5 km^2. This area comprises historical urban tissue, a market, open spaces, and a park. This area of Qazvin also provides a variety of places, turning the places into a varied and multifunctional space for pedestrians.

The streets in the area selected in Porto sum 47 km, while in Qazvin sum 45 km. However, 64% of the streets in Qazvin has a length less than 120 m, while in Porto this percentage decreases to 46% (Fig. 15.2). These differences are directly related with the urban morphologies. Qazvin has a more polycentric structure defined by traditional neighbourhoods and main squares, containing many cul-de-sacs and small streets. The area selected in Porto is defined by only one centre ('Baixa') where the political, the religious and the main economic activities were installed.

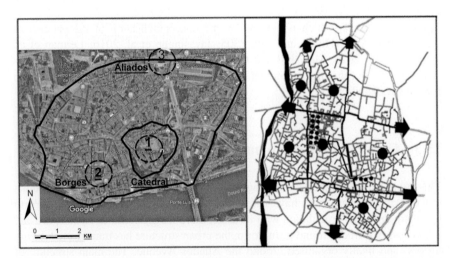

Fig. 15.1 Urban structure and process of urban growth between the 12th century and 16th century in Porto (left) and Qazvin (right). *Source* Google Map, 2019

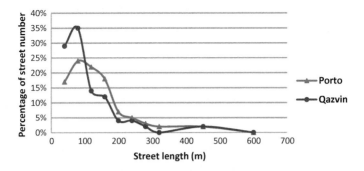

Fig. 15.2 Percentage of streets based on length (Porto and Qazvin). *Source* Authors

15.4 Methodology

The methodological procedures adopted in the study will be described according to the following order (Fig. 15.3): (i) Criteria selection; (ii) Description of the data collection; (iii) weights assignment and aggregation method; (iv) Connectivity Evaluation; (v) Assessment of the pedestrian network; and (vi) Obtain the cohesive pedestrian network.

As shown in Fig. 15.3, the first step of the work consisted of selecting the criteria with impact on walking. This work was done by making a literature review and by consulting a panel of experts. As described on Sect. 15.2, the criteria selected with impact on walking were the physical environment, urban functions, accessibility and natural environment. Each criteria was subdivided in sub-criteria as shown in Table 15.1.

The second step of the work was the validation of this initial list of criteria by a panel of experts in urban planning from Porto and Qazvin. Experts were involved in the study through surveys performed between July and September 2015. A total of 41 experts from Porto and 45 experts from Qazvin participated in the study. The first part of the survey consisted of selecting/validating the list of criteria and sub-criteria presented in Table 15.1. The survey was prepared to allowed the inclusion of new criteria/sub-criteria. Nonetheless, experts confirmed the options given in the list as only very few comments were made. For instance, safety was not highlighted as a critical criterion for walking demonstrating that this factor is not

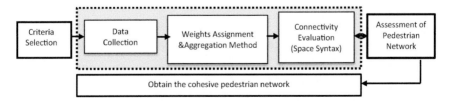

Fig. 15.3 Steps followed to define a pedestrian network assessment model. *Source* Authors

Table 15.1 Criteria and sub-criteria used to assess the conditions provided to pedestrians

Criteria	Sub-criteria
Physical environment	Terrain slope
	Human scale
	Visual dimension
Urban function	Land use
	Population density
Accessibility	Public transportation service
	Intelligent transportation system(ITS)
Natural environment	Green space
	Microclimatic conditions (Temperature)

Source Authors

relevant in Porto and Qazvin. Based on this survey, the four criteria and the nine sub-criteria included in Table 15.1 were selected for making the PNA proposed in this study.

The survey addressed to experts was also used for defining the relative importance of the several criteria and sub-criteria. In this step of the work, experts were invited to assign weights to each criteria/sub-criteria according to their view. As criteria were valued differently, it was necessary to normalise the values. Criteria were normalised by using the sigmoidal function presented in Eq. 15.1.

$$f(x_i) = \begin{cases} x = x_{\min} = 0 \rightarrow f(x_i) = 0 \\ x_{\min} < x_i < x_{\max} \rightarrow f(x_i) = \frac{x - x_{\min}}{x_{\max} - x_{\min}} \\ x = x_{\max} \rightarrow f(x_i) = 1 \end{cases} \quad (15.1)$$

where,

X_i—Element of the network ($i = 1, 2,..., n$)

X—X_i of elements of the network

After normalising the sub-criteria, the next step was the aggregation of the relative values by using the weighted linear combination (WLC) method for sub-criteria (2) and criteria (3). Therefore, each standardised factor was multiplied by the respective weight. The process was adopted for obtaining the sub-criteria (Eq. 15.2) and the criteria (Eq. 15.3) weights.

$$j = 1, 2, \ldots, 9 \; S_k(x_i) = \sum_{i=0}^{n} f_j^k(x_i) w_{kj} \quad (15.2)$$

$S_k(x_i)$—Assessment of element X_i for all the sub-criteria j of criteria k

$$k = 1, 2, \ldots, 4 \; T(x_i) = \sum_{i=0}^{n} S_k(x_i) w_k \quad (15.3)$$

Table 15.2 Weight of criteria and sub-criteria in Porto and Qazvin

Criterion	Weight	Sub criterion	Weight
Porto			
Physical environment	0.235	Terrain slope	0.290
		Human scale	0.295
		Visual dimension	0.415
Urban Function	0.253	Land use	0.875
		Population density	0.125
Accessibility	0.224	Public transportation service	0.50
		Intelligent transportation system (ITS)	0.50
Natural environment	0.288	Green space	0.667
		Microclimatic conditions (temperature)	0.333
Qazvin			
Physical environment	0.179	Terrain slope	0.273
		Human scale	0.228
		Visual dimension	0.497
Urban function	0.199	Land use	0.264
		Population density	0.736
Accessibility	0.310	Public transportation service	0.576
		Intelligent transportation system (ITS)	0.424
Natural environment	0.312	Green space	0.565
		Microclimatic conditions (temperature)	0.435

Source Authors

$T(x_i)$—Assessment of all the criteria k for the element X_i of the network

Converting the experts' opinions into values was done by using a pairwise comparison matrix. Therefore, the sum of all weights obtained using this method is equal to one. Table 15.2 presents the weights assigned by the experts of Porto and Qazvin. As shown in Table 15.2, the sub-criteria and criteria weights were calculated by performing different pairs of combinations. The criteria more valued by the experts from Porto were natural environment and urban function. In turn, the experts of Qazvin also ranked first the natural environment, followed by accessibility.

The next step was assessing the street network connectivity. Space syntax and more particularly the DepthmapX software was used to analyse street connectivity. Higher space syntax values correspond to streets with many connections (nodes) and vice versa. The analysis was performed by using the Eq. 15.4.

$$C_i = \sum_k R_{ik}, \qquad (15.4)$$

where

R_{ik} presents the direct link between i and k.

In classical theory, C_i can positively quantify the permeability of i together with its aggregation degree; Based on the graph theory, the connectivity of a node can be defined as the number of other nodes directly connected to it. The analysis performed with space syntax shows a street connectivity ranging from 1 to 14 in Porto and from 1 to 21 in Qazvin. Higher space syntax values correspond to streets with many connections (nodes) and vice versa. These values were then normalised between 0.0 and 1.0 by fuzzy logic and inserted in the GIS database.

The last step of the work consisted of aggregating the street connectivity with the multi-criteria values. A WLC was calculated again to obtain the final scores by using a weight of 0.5 for the multi-criteria evaluation and 0.5 for street connectivity, as suggested by the experts. The result is a global assessment of the conditions provided to pedestrians (PNA), by considering characteristics of the physical and natural environment and street connectivity.

Finally, a quadrant chart method was proposed to compare the conditions provided by the two cities. As highlighted by Oh and Jeong (2007) and by Zhou et al. (2015), quadrant chart method is a useful tool for setting up items that shares common attributes by representing data in separate quadrants.

15.5 Results

This section presents the results obtained with the described method for Porto and Qazvin. In Porto, the sub-criteria assigned with higher values were population density, followed by the proximity to public transport and microclimatic conditions. In Qazvin, land use was the criterion more relevant for the experts, while the less important was population density. In turn, for the Porto experts, the less relevant criterion was human scale. In terms of criteria, which weights were aggregated by using a pairwise comparison matrix, the natural environment was the criterion more important for the experts of both cities. Such finding is in line with other studies demonstrating that the conditions of the natural environment (thermal comfort, shadow, natural light, etc.) are very important for pedestrians (Choi et al. 2016; Jayasinghe et al. 2016).

The ranking combining the four criteria was obtained by overlaying the four layers by using the weights presented in Table 15.2. This evaluation was implemented in GIS. Figure 15.4 shows the classification of the streets of the selected areas of Porto (4a) and Qazvin (4b) according to the four criteria. The highest classifications obtained were 0.71 in Porto and 0.75 in Qazvin. The average of the values obtained in both cities were also very similar: 0.387 in Porto and 0.393 in Qazvin. Results also showed that 52% of the streets in Porto and 56% of the streets in Qazvin ranked above the mentioned average values. In other words, more than half of the streets analysed in both cities score above average.

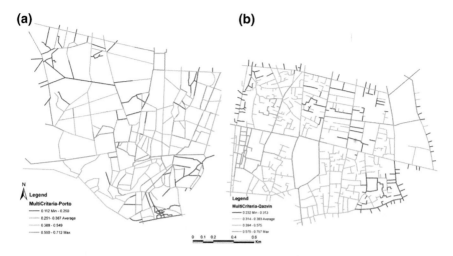

Fig. 15.4 Streets ranking considering the multi-criteria analysis in Porto (**a**) and in Qazvin (**b**). *Source* Authors

As described in the methodology, the street network connectivity analysis was applied by using Space Syntax and the results for both cities are presented in Fig. 15.5. The streets of Porto with higher levels of connectivity are in the central area especially in Bolhão and Avenida dos Aliados. The lowest levels of connectivity were found in the western area which includes the Boavista area, and also in some parts of the historical centre. Street connectivity in Porto obtained an average

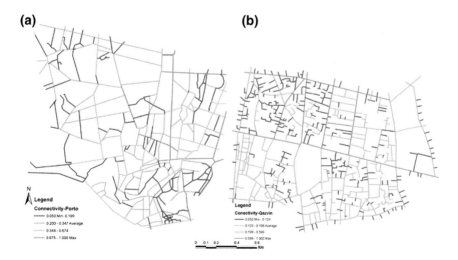

Fig. 15.5 Streets ranking based on the street network connectivity in Porto (**a**) and in Qazvin (**b**). *Source* Authors

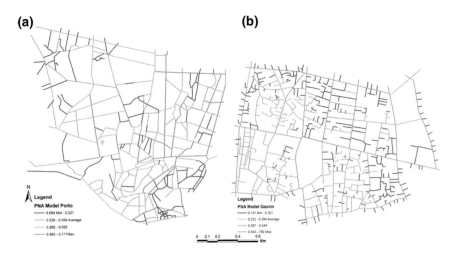

Fig. 15.6 Results of the PNA model in Porto (**a**) and in Qazvin (**b**). *Source* Authors

value of 0.347 and 39% of the street length had a connectivity ranking above the average. In Qazvin, the streets with high connectivity were Naderi, Ferdowsi, Bazaar and Bu Ali Sina. Old neighbourhoods surround these streets. In turn, the streets with lower connectivity are located in Southern part of Qazvin, which is more recent. Street connectivity in Qazvin obtained an average value of 0.198, but 41% of the street length had a connectivity ranking above the average.

The last stage of PNA was combining the Multi-Criteria Analysis (MCA) with the street network connectivity. The results are presented in Fig. 15.6. The comparison with Fig. 15.4 shows that the inclusion of the connectivity increased the ranking obtained by the streets. In Fig. 15.6, the highest value increased from 0.71 to 0.77 in Porto; while in Qazvin, the highest value improved from 0.75 to 0.79. The percentage of streets ranking above the average values obtained decreased in Porto from 52% to 49%, while in Qazvin the values keep almost unalterable.

15.6 The Impact of Urban Morphology on the Pedestrian Network

As described in the methodology, a quadrant chart method was proposed to compare the conditions provided to pedestrians by the two cities. The quadrant method is considered a helpful decision making tool by representing data on two axes (X and Y).

In this study, the horizontal axis represents the MCA index, whereas the vertical axis shows the SNC index. In the chart, each point corresponds to the streets assessed in Porto and Qazvin (Fig. 15.7). In the quadrant chart, values of MCA and

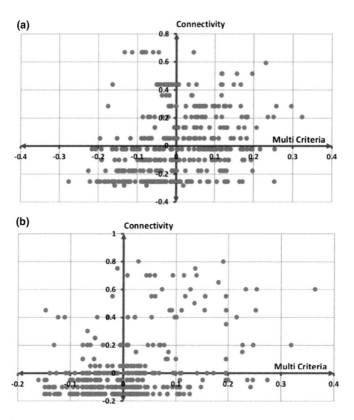

Fig. 15.7 Quadrant chart according to data average for Porto (**a**) and Qazvin (**b**). *Source* Authors

SNC can be greater or lesser than zero and the four following main situations can be found:

First quadrant (high MCA and SNC): streets located on this quadrant have positive values (MCA > 0 and SNC > 0) considering the criteria analysed and street connectivity. Streets located on this quadrant have the highest levels of walkability.

Second quadrant (low MCA and high SNC): streets located on this quadrant have positive values in terms of connectivity (SNC > 0) but negative values in terms of criteria (MCA < 0). The walkability of these streets could be enhanced by taking actions for improving the criteria with poor performance.

Third quadrant (low MCA and SNC): streets located on this quadrant have negative values (MCA < 0 and SNC < 0) considering the criteria analysed and street connectivity. Streets located on this quadrant have the lowest levels of walkability.

Fourth quadrant (high MCA and low SNC): streets located on this quadrant have positive values in terms of criteria evaluation (MCA > 0) but negative values in terms of street connectivity (SNC < 0). The walkability of these streets could be enhanced by taking actions for improving street connectivity.

The aforementioned classification provided by the quadrant chart was used for ranking the streets according to their potential to define a network of pedestrian streets. Thus, the streets were organised in the following three-level scale:

- The first level refers to the streets that have the highest potential to make a pedestrian network. This includes the streets included in the first quadrant.
- The second level includes streets that have the potential to make part of a pedestrian network. These streets require actions to improve their connectivity or specific conditions provided to pedestrians. This level includes the streets located in the second and in the fourth quadrants.
- The third level includes the streets that have the lowest potential to make a pedestrian network. This includes the streets located in the third quadrant.

For a better comparison of the conditions provided by the streets of Porto and Qazvin, the total length of the streets located in each quadrant was measured. Results of this assessment are presented in Table 15.3. The length of the streets analysed in the two cities was similar: about 47 km in Porto and 45 km in Qazvin. Considering the street length, Qazvin had 41% of the streets located in the first quadrant, 32% in the second and fourth quadrants and 28% in the third quadrant. In turn, Porto had 27% of the streets in the first quadrant, 36% in the second and fourth quadrants and 37% in the third quadrant. As the first quadrant comprises the streets providing better conditions for pedestrians and the highest potential to make a

Table 15.3 Multi-criteria (X) and street network connectivity (Y) in Porto and Qazvin from quadrant chart

Porto city	Quadrant	Street length (m)	Percentage
MCA > 0.387 and SNC > 0.330	1 (X > 0 and Y > 0)	12558.10	27
MCA <= 0.387 and SNC > 0.330	2 (X < 0 and Y > 0)	11137.34	24
MCA <= 0.387 and SNC <= 0.330	3 (X < 0 and Y < 0)	17355.02	37
MCA > 0.387 and SNC <= 0.330	4 (X > 0 and Y < 0)	5882.26	12
Total length of Porto		46932.74	100
Qazvin city	Quadrant	Street length (m)	Percentage
X > 0.392 and Y > 0.198	1 (X > 0 and Y > 0)	18087.20	41
X <= 0.392 and Y > 0.198	2 (X < 0 and Y > 0)	7019.98	16
X <= 0.392 and Y <= 0.198	3 (X < 0 and Y < 0)	12400.02	28
X > 0.392 and Y <= 0.198	4 (X > 0 and Y < 0)	7042.90	16
Total length of Qazvin		44544.32	100

Source Authors

pedestrian network, it can be concluded that Qazvin exhibits general better conditions to pedestrians than Porto. In this last city, the length of streets included in the worst quadrant (3) and in the intermediate quadrants (2 and 4) is much more representative.

A global overview of the streets ranked in the first level (streets on the first quadrant) as well as in the second level (streets on the second and fourth quadrants) are respectively presented in Figs. 15.8 and 15.9. In Porto, the streets classified in the first level are mainly located around Praça da Liberdade, Avenida dos Aliados, S. Bento station, Mouzinho da Silveira. They broadly define the central area of the city called 'Baixa' which corresponds to the oldest main centrality of the city. In Qazvin the situation is different, it is evident that there are not only more streets classified in this level, but also a higher distribution among several centres. Figure 15.8b shown the streets of Qazvin have with higher levels of connectivity includes the main streets named to Nadri, Ferdowsi, Bazar and Bu Ali Sina streets. Also, those streets have surrounded the ancient neighbourhoods as the boundary. Regarding the streets classified in the second level, the conclusions are identical. In Qazvin, the streets are arranged around in three neighbourhoods that named Drab-Kooshek, Bolaghi, Sar-kocheh reyhan and tradition market (Bazaar). In the case of Porto, there are much more streets classified in the second level and they are more widespread by the selected area. The Western (more recent) part of the city, as well as part of Ribeira, characterised by narrowed and hilly streets, are the areas of Porto providing less walkable conditions.

The different performance obtained in both cities can be attributed to the different urban morphologies. Qazvin has a more polycentric structure due to the traditional neighbourhoods arranged around centres (mosque, bazaar). In Porto, the urban morphology is more concentrated around 'Baixa', the traditional political and

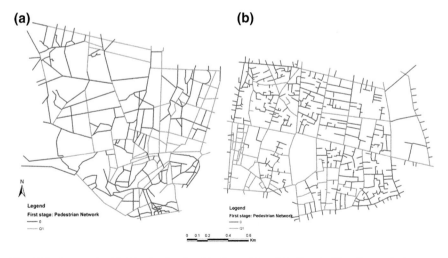

Fig. 15.8 Streets classified in the first level in Porto (**a**) and in Qazvin (**b**). *Source* Authors

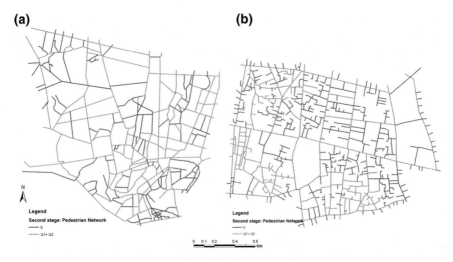

Fig. 15.9 Streets classified in the second level in Porto (**a**) and in Qazvin (**b**). *Source* Authors

economic centre. The PNA model also shows that a network of pedestrian streets is particularly missing in Porto, where streets providing good conditions are well-connected only cover a small area. The PNA model was also useful for identifying a group of streets with potential to improve the pedestrian network—the second level streets. By defining actions to improve the walkability and/or connectivity of these streets, these findings could be fundamental for obtaining a broader and well-connected network of pedestrian streets. Therefore, the model described in this paper as potential to support urban planning decisions aiming at improving walkability and sustainability in cities.

15.7 Conclusion

The main aim of this research was to compare the conditions provided to pedestrians in two cities with different urban morphologies: Qazvin (Iran) and Porto (Portugal). It proposed a pedestrian network assessment (PNA) model considering multi-criteria and street connectivity analyses. A quadrant chart method was also proposed to analyse and compare the results obtained in the two cities. Results show that Qazvin provides more walkable streets considering the criteria analysed and street connectivity. Moreover, the pedestrian streets are more organised and connected in a network in Qazvin than in Porto. By identifying the problems affecting walkability and the level of connectivity, the PNA model has a potential to help planners and decision-makers in designing more walkable cities. The described model can also potentially be replicated in other cities in terms of improving walkability and in promoting sustainable urban mobility.

In addition, the urban structure determines how serviceable and flexible an urban area is and how well it integrates into its surroundings. The urban structure contributes to both the function and feel of an area and creates a sense of place. A well-functioning urban structure connects neighbourhoods, where activity centres are within a convenient walking distance. Regarding type of urban spatial structures, the polycentric urban structure is an effective way to reduce pressure on transport to the city centre and influences the possibility of walking. Research results are attributed to the polycentric structure of Qazvin, causing a strong network of walkable streets in Qazvin.

Nonetheless, some aspects could be improved in future works. It will be particularly important to support the weighting process in a larger and representative sample. A combined system involving not only a group of experts but also the residents' opinions could be useful to strengthen the robustness of the approach. Finally, the model will present a more accurate assessment from the walkable streets in a central area of a city to identify the potential for a pedestrian network and to improve mobility.

References

Badland H, White M, MacAulay G, Eagleson S, Mavoa S, Pettit C, Corti B (2013) Using simple agent-based modeling to inform and enhance neighborhood walkability. Int J Health Geogr 12–58

Bahrainy H, Khosravi H (2013) The impact of urban design features and qualities on walkability and health in under-construction environments: the case of Hashtgerd New Town in Iran. Cities J 31:17–28

Buccolieri R, Salizzoni P, Soulhac L, Garbero V, Sabatino S (2015) The breathability of compact cities. Urban Climate 13:73–93

Cervero R, Kockelman K (1997) Travel demand and the 3Ds: density. Diversity Design Transp Res D 2:199–219

Choi W, Ranasinghe D, Bunavage K, DeShazo JR, Wu L, Seguel R, ... Paulson SE (2016) The effects of the built environment, traffic patterns, and micrometeorology on street level ultrafine particle concentrations at a block scale:results from multiple urban sites. Sci Total Environ 553:474–485. https://doi.org/10.1016/j.scitotenv.2016.02.083

Cubukcu E, Hepguzel B, Onder Z, Tumer B (2015) Active living for sustainable future: a model to measure "walk scores" via geographic information systems. Proc—Social Behav Sci 168:229–237

Dallman A, Di Sabatino S, Fernando HJS (2013) Flow and turbulence in an industrial/suburban roughness canopy. Environ Earth Sci J 13:279–307

Evans G (2001) Cultural planning: an urban renaissance?. Routledge, London

Farrell T (2000) Urban regeneration through cultural master planning. In: Rose JBaM (ed) Rotterdam: urban lifestyles: spaces, places, people

Fernando HJS, Zajic D, Di Sabatino S, Dimitrova R, Hedquist B, Dallman A (2010) Flow, turbulence, and pollutant dispersion in urban atmospheres. Physics 22:1–20

Ferreira IA, Johansson M, Sternudd C, Fornara F (2016) Transport walking in urban neighbourhoods—impact of perceived neighbourhood qualities and emotional relationship. Landscape Urban Plann 150:60–69

Forsyth A, Oakes M, Lee B, Schmitz K (2009) The built environment, walking, and physical activity: is the environment more important to some people than others? Transp Res Part D: Transport Environ 14(1):42–49

Gilderbloom J, Riggs W, Meares W (2015) Does walkability matter? An examination of walkability's impact on housing values, foreclosures and crime. Cities 42:13–24

Gospodini A (2006) Portraying, classifying and understanding the emerging landscapes in the post-industrial city. Cities J 23:311–330

Grecu V, Morar T (2013) A decision support system for improving pedestrian accessibility in neighborhoods. Proc Soc Behav Sci 92:588–593

Green CG, Klein EG (2011) Promoting active transportation as a partnership between urban planning and public health: the Columbus healthy places program. Public Health Rep 126(1):41–49

Havard C, Willis A (2012) Effects of installing a marked crosswalk on road crossing behavior and perceptions of the environment. Transp Res Part F 15:249–260

Hillier B, Hanson J (1998) Space syntax as a research programme. Urban Morphol 2:108–110

Hillier B, Iida S (2005) Network effects and psychological effects: a theory of urban movement. Paper presented at the Fifth international space syntax symposium, Delft, Netherland

Hillier B, Perm A, Hanson J, Grajewski T, Xu J (1993) Natural movement: or configuration and attraction in urban pedestrian movement. Environ Plan 19:29–66

Jabbari M, Fonseca F, Ramos R (2017) Combining multi-criteria and space syntax analysis to assess a pedestrian network: the case of Oporto. J Urban Des 23(1):23–41. https://doi.org/10.1080/13574809.2017.1343087

Jamei E, Rajagopalan P (2017) Urban development and pedestrian thermal comfort in Melbourne. Sol Energy 144:681–698. https://doi.org/10.1016/j.solener.2017.01.023

Jayasinghe A, Sano K, Kasemsri R, Nishiuchi H (2016) Travelers' route choice: comparing relative importance of metric, topological and geometric distance. Proc Eng 142:18–25. https://doi.org/10.1016/j.proeng.2016.02.008

Jeong SK, Banyn YU (2016) A point-based angular analysis model for identifying attributes of spaces at nodes in street networks. Physica A: Stat Mech Appl 450:71–84

Jiang B, Liu C (2009) Street-based topological representations and analyses for predicting traffic flow in GIS. Int J Geogr Inf Sci 23(9):1119–1137. https://doi.org/10.1080/13658810701690448

Koh P, Wong Y (2013) Influence of infrastructural compatibility factors on walking and cycling route choices. J Environ Psychol 36:202–213

Lamíquiz PJ, López-Domínguez J (2015) Effects of built environment on walking at the neighbourhood scale. A new role for street networks by modelling their configurational accessibility? Transp Res Part A 74:148–163

Lerman Y, Omer I (2016) Urban area types and spatial distribution of pedestrians: lessons from Tel Aviv. Comput Environ Urban Syst 55:11–23

Lerman Y, Rofè Y, Omer I (2014) Using space syntax to model pedestrian movement in urban transportation planning. Geograph Anal 46:392–410

Li Y, Xiao L, Ye Y, Xu W, Law A (2016) Understanding tourist space at a historic site through space syntax analysis: the case of Gulan gyu, China. J Tourism Manag 52:30–43

Lundberg B, Weber J (2014) Non-motorized transport and university populations: an analysis of connectivity and network perceptions. J Transp Geogr 39:165–178

Marquet O, Miralles-Guasch C (2016) City of motorcycles. On how objective and subjective factors are behind the rise of two-wheeled mobility in Barcelona. Transp Policy 52:37–45. https://doi.org/10.1016/j.tranpol.2016.07.002

Martinelli L, Battisti A, Matzarakis A (2015) Multicriteria analysis model for urban open space renovation: an application for Rome. Sustain Cities Soc 14:10–20. https://doi.org/10.1016/j.scs.2014.07.002

McCahill C, Garrick NW (2008) The applicability of space syntax to bicycle facility planning. J Transp Res Board 2074:46–51

Mehta V (2008) Walkable streets: pedestrian behavior, perceptions and attitudes. J Urbanism: Int Res Placemaking Urban Sustain 1(3):217–245

Moura F, Cambra P, Gonçalves AB (2017) Measuring walkability for distinct pedestrian groups with a participatory assessment method: a case study in Lisbon. Landscape Urban Plann 157:282–296. https://doi.org/10.1016/j.landurbplan.2016.07.002

Nasir M, Lim C, Nahavandi S, Creighton D (2014) A genetic fuzzy system to model pedestrian walking path in a built environment. Simul Model Pract Theory 45:18–34

Oh K, Jeong S (2007) Assessing the spatial distribution of urban parks using GIS. Landscape Urban Plann 82(1–2):25–32. https://doi.org/10.1016/j.landurbplan.2007.01.014

Önder DE, Gigi Y (2010) Reading urban spaces by the space-syntax method: a proposal for the South Haliç Region. 27:260–271

Panagopoulos T, Duque J, Dan M (2016) Urban planning with respect to environmental quality and human well-being. Environ Pollut 208:137–144

Peiravian F, Derrible S, Ijaz F (2014) Development and application of the Pedestrian Environment Index (PEI). J Transp Geography (39):73–84

Peponis J, Hadjinikolaou E, Livieratos C, Fatouros DA (1989) The spatial core of urban culture. In: Ekistics. Athens Center of Ekistics of the Athens Technological Organization, 1/1989, pp 43–55

Penn A, Hillier B, Banister D, Xu j (1998) Configurational modelling of urban movement networks. Environ Plann B: Plann Des 25(1):59–84

Pourjafar M, Amini M, Hatami Varzaneh E, Mahdavinejad M (2014) Role of bazaars as a unifying factor in traditional cities of Iran: The Isfahan bazaar. Front Architect Res J 3:10–19

Socharoentum M, Karimi HA (2016) Multi-modal transportation with multi-criteria walking (MMT-MCW): personalized route recommender. Comput Environ Urban Syst 55:44–54. https://doi.org/10.1016/j.compenvurbsys.2015.10.005

Tavassoli M (2016) Urban structure in Islamic territories urban structure in hot arid environments: strategies for sustainable development. Springer International Publishing, Cham, pp 11–18

Zadeh LA (1978) Fuzzy sets as a basis for a theory of possibility. J Fuzzy Sets Syst 1:3–28

Zhou J, Zhang X, Shen L (2015) Urbanization bubble: four quadrants measurement model. Cities 46:8–15. https://doi.org/10.1016/j.cities.2015.04.007

Chapter 16
Challenges of Participatory Urban Design: Suggestions for Socially Rooted Problems in Sang-e Siah, Shiraz

Elham Souri, Jahanshah Pakzad and Hooman Foroughmand Arabi

Abstract A number of studies suggest that a poor engagement with stakeholders and problematic procedures of implementation are the key reasons for failures of many urban design projects in Iran. The international literature suggests that engaging with a wide range of social groups has a crucial role in delivering successful urban projects. However, there are many obstacles for participation. This chapter aims to identify the main constrains of public participation in urban projects by focusing on the case of Sang-e Siah in the historical part of the city of Shiraz in Iran. Encompassing numerous historical buildings, Sang-e Siah is one of the most important historical districts of Shiraz, due in part to its residents who have been living in this neighbourhood for generations. In fact, this area is an important case as it embodies a wide range of urban issues which are common in many other Iranian cities. The chapter commences by introducing a problematic participatory urban design project for the area. This is followed by identifying the main social obstacles found through qualitative interviews (with 20 residents of the neighbourhood) and meetings with authorities (including the municipality of Shiraz and other organisations responsible for providing and maintaining urban infrastructures). Through the interviews, the residents' perceptions and concerns about the neighbourhood were found to be important. The study concludes that there are two major burdens to public participation in the urban project: (a) mental constraints of both residents and authorities, embodied in misunderstandings of the participatory approaches (b) the challenging socio-economic conditions of the community embodied in high rates of social problems. These findings can be better explained in the light of the history of the Iranian society, which points at culturally produced, mental burdens against engagement with communal activities. Participatory

---

E. Souri (✉) · J. Pakzad
Shahid Beheshti University, SBU, Tehran, Iran
e-mail: elham.souri@gmail.com

J. Pakzad
e-mail: jahanshahpakzad.u@gmail.com

H. Foroughmand Arabi
University of West England, Bristol, UK
e-mail: hooman.araabi@ucl.ac.uk

© Springer Nature Switzerland AG 2020
F. F. Arefian and S. H. I. Moeini (eds.), *Urban Heritage Along the Silk Roads*,
The Urban Book Series, https://doi.org/10.1007/978-3-030-22762-3_16

activities in particular face deep-rooted issues requiring long-term supportive programmes such as educational and cultural initiatives, some of which recommended here. This study offers insights into participatory urban design in Iranian historical districts. As a long-term policy, educational and cultural changes are discussed. The long-term policy follows enhancing socio-economic characteristics of the communities through delivering environmental qualities, as a mid-term policy.

Keywords Participatory urban design · Social problems · Sang-e Siah · Shiraz

16.1 Introduction

The quality of the built environments, as one concern of urban designers, can be affected by a vast variety of forces called 'actors'. It is noteworthy mentioning that residents, property owners, shopkeepers, users, politicians and city authorities are critical and powerful parties in determining the quality of built environments.

Although the international literature suggests that engaging with a wide range of social groups has a crucial role in the delivery of successful urban projects, many Iranian studies acknowledge and draw attention to the consequences of poor engagement with stakeholders and problematic procedures of implementation as the key reasons for failures of many urban design projects in Iran (Shahid Beheshti University 2016, Saghafi et al. 2016, and Safavi 2011). Many studies have been undertaken to identify the primary targets and steps for participatory design. What is less known, however, is the practical constraints especially in historical parts of the city (Pakzad 2002 and 2004).

This chapter tries to demonstrate the main constrains of public participation in urban regeneration projects in the case of Sang-e Siah in the historical part of the City of Shiraz in Iran. The participatory design approach is described briefly in Sect. 16.2 below, followed by an introduction to the district and its main actors (Sects. 16.3.1 and 16.3.2), a discussion in Sect. 16.3.4 about the constrains of the public participation, and then a conclusion based on findings.

16.2 Participatory Urban Design

Participatory design (PD), is an approach in which all stakeholders are invited to be involved and cooperate with designers and researchers in the planning and designing process. Potentially, they participate during several stages: problem definition, visioning, surveying, studying and evaluating as well as creating and developing design ideas (Tan and Szebeko 2009; Sanders and Stappers 2008). Traditional urban design projects typically involve the client and the professional, whilst participatory urban design, involves members of the broader community, from the users who are directly affected by the design, to the local business owners, who are also recognised as legitimate stakeholders to have a say about the project. Therefore, participatory design is 'a project of entanglement of many different design elements' and the designer's role becomes that of a collaborator in the construction of 'a meaningful potentially controversial gathering, for and with the participants in the projects' (Manzini and Rozzo 2011). As Lee (2008) discusses, the scope of stakeholders' (or users') involvement in design process can range from being passively informed of a project's development, to actively sharing roles and responsibilities in decision-making.

In planning and designing process of Sang-e Siah, participatory approach was the primary issue, because:

(a) It provides the working team with the chance of being assured of consequences of each decision based on real findings rather than making assumptions regarding actual users' and beneficiaries' needs, interests and expectations. They could understand and predict the success rate of different ideas and consequently users' reactions.
(b) Moreover, participatory approach creates an appropriate platform for developing a community members' sense of place attachment. In other words, Individuals consider themselves the main owners of design outcomes and voluntarily maintain it when they are involved in design process. This was echoed in Sang-e Siah planning process as there was a belief that people are able to present realistic, innovative solutions, and that instead of prescribing solutions for people, planners can facilitate them with people.
(c) Also, the main focus is on strengths. Although there are restrictions attached to participatory design, there is an opportunity to take advantage of the stakeholders' abilities to achieve a more realistic approach in design process.
(d) Furthermore, participatory approach provides an appropriate platform for re-educating general public through expanding all participants' urban design knowledge and skills. Opening critical dialogue amongst various members is a contributing factor for exploring creative and practical strategies.

16.3 Sang-e Siah District Design Through Participatory Approach

With the aim of urban regeneration of Sang-e Siah historical neighbourhood the current project was defined and conducted by Urban Development and Revitalisation Organisation (UDRO) as the client in 2008 and lasted until 2010. As mentioned, since participatory design was at the core of the design team's approach, they made an effort to enhance public participation from the outset. This resulted in restrictions as well as possibilities: each with their own lessons.

16.3.1 The Location

Sang-e Siah is actually one of the most important historical pathways in Shiraz with a length of 810 metres passing through several historical neighbourhoods (e.g. Sang-e Siah, Darb-e Masjed, Sar Bagh) providing a religious-cultural centre for the city from past to present (Fig. 16.1). A combination of various factors accounts for why this pathway is very significant:

- The inclusion in the area of more than 30 historical cultural constructions such as Mooshir mosque, Bibi Dokhtaran Holy Sepulchre, an Ilkhani mansion and garden, Forough-al-Molk School, Armenian Bazaar (Figs. 16.2 and 16.3)
- The co-presence in the area of people of various faiths.
- The main connection through the pathway between two important roads to the north and south of the historical district, namely Sibouyeh and Pirouzi Avenues.
- The multigenerational history of people of fame having lived in this neighbourhood (e.g. those from Owji and Mohandesi families).

16.3.2 The Main Urban Actors in the Neighbourhood

The Urban Design Group (UDG) took the view that everyone acting in this environment is an urban designer because the decisions he or she makes affect the quality of urban space (Linden and Billingham 1998, p. 40). Thus, just as Carmona et al. (2003) have argued, all those who take a decision about the urban environment which shapes the place are urban designers of sorts, and should be involved in the design process. In this project the primary actors were recognised as:

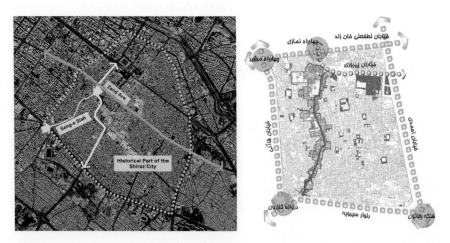

Fig. 16.1 Location of the Sang-e Siah pathway in historical part of the Shiraz

Fig. 16.2 Mooshir mosque

Fig. 16.3 Bi-Bi Dokhtaran holy sepulchre

a. Residents, tradespeople and tourists as real users of the spaces,
b. The Municipality of the Historical Fabric, Fars Cultural Heritage, and the Handicraft & Tourism Organisation, officially as the main authorities responsible for management and maintenance of this district.
c. Other organisations responsible for providing and maintaining urban infrastructures.

Although all actors were studied, the focus was particularly on residents and local authorities.

16.3.3 Research Method

The method used in this study was based on a qualitative survey. The purpose of using this method was achieving an in-depth understanding of the circumstances, needs, expectations and attitudes of residents as well as local authorities.

Research participants consisted of 20 residents of Sang-e Siah neighbourhood and 10 city authorities from the Municipality of Shiraz and other organisations responsible for providing and maintaining urban infrastructures.

16.3.4 Main Constrains of the Public Participation in the Neighbourhood

Through the interviews, the residents' perceptions and concerns of the neighbourhood were found to be paramount factors. The data analysis indicates that there are two major burdens against public participation in urban project: (a) mental constraints of both residents and authorities, embodied in the misunderstandings of the participatory approaches, and (b) the challenging socio-economic conditions of the community embodied in high rates of social problems.

16.3.4.1 Mental Constraints

This was found in both residents and local authorities:

Residents

The data analysis based on interview indicates a combination of various factors' accounts for mental constraints of residents' participation, including:

(a) Not only do all participants consider public spaces as governmental and outsider spaces, but they also do not make any efforts towards the management and maintenance of these public spaces. To be precise, they do not consider themselves as members of urban management systems.
(b) A distrust in teamwork along with a view about urban activities as political action are among notable reasons why inhabitants tend not to participate in design. Actually, they consider local authorities such as the Municipality as the main bodies responsible for regeneration and management of open spaces.
(c) The lack of teamwork experience has led to authoritarianism among occupants; sometimes they look for a hero who can save them, occasionally they cannot tolerate an opposite point of view.
(d) Although urban design is a long-term activity, they tend to expect quick outcomes and find themselves frustrated with lengthy procedures.

Local Authorities and City Managers

Mental constraints of local authorities to participatory design include:

(a) Local authorities traditionally consider general public as an uneducated and illiterate mass who need care regarding planning for the future. They unconsciously draw a line between themselves and the public and assert a very special position for themselves in the planning system. They view the people's

environment as unhealthy and their strategies and actions as healing. This has led to an implicit view that common people are not entitled to collaborate with local authorities in urban design procedure. There are several indications that local authorities do not welcome expanding residents' understanding of their rights, as the presence of this gap is found beneficial to them.
(b) There is generally an organisational prejudice among local authorities. Not only do they distrust the general public, they also believe other related institutions are less reliable than theirs. Cherishing their own organisational objectives as holy, coordinating and collaborating with other organisations of different agenda has proved tough for local authorities' management.
(c) The usually short lifespan of city managerial jobs in Iran leaves little room for well-thought, long-term decision-making. The patience required for participatory design methods is, therefore, at odds with this short-termism and its passion for quick, visible outcomes.
(d) Due to the necessity of demonstrating their competency, local authorities usually consider urban design projects as platforms for creating a design masterpiece, which is frequently at odds with the urgency and economic awareness they are keen to demonstrate.
(e) Local authorities' perception of public participation is mostly that of residents' financial contribution rather than their real collaboration in decision-making.
(f) Last but not least, is local authorities' lack of disbelief in the key role of occupant involvement in public acceptance, success, maintenance and management of urban design projects.

16.3.4.2 Socio-economic Conditions of the Community

Obtaining a deep understanding of socio-economic conditions of the project context is one of primary essentials in participatory design. About 10% of Shiraz's population lives in the historical part of the city, whilst the population growth in such areas does not reflect that of the city as a whole. In other words, the statistics demonstrate a gradual reduction fall in the population of the historical district. The researchers' observations and surveys indicate that the native population of the historical part of Shiraz, especially Sang-e Siah, is increasingly being replaced with new residents. Many studies have been undertaken to identify the targets destinations of local population emigration from this district. The image of Sang-e Siah district has increasingly worsened in terms of its physical, socio-economic and cultural aspects, and these are recognised as the most significant reasons for the district's deterioration.

In other words, owing to the fact that this old district has been less than responsive to modern life requirements, indigenous and original residents have gradually been wiped out. A high depreciation rate of old buildings along with the lack of funds and expertise in restoration, and accessibility and movement problems have all contributed to the failure of old buildings to meet residents' needs, and their gradual replacement with low-income and non-local individuals, especially Afghans who move to this region in search for jobs and affordable houses.

Obviously, an appropriate insight about needs, tastes, and tendencies of current residents in Sang-e Siah plays a crucial role in finding the participatory design difficulties and the way of combating the restrictions.

The main dwellers can be categorised into four groups: young and single male immigrants, families who have no other choice, families who choose to live in this region, and finally drug addicts and dealers. Collective residence of young and single immigrant men not only accelerate the old constructions' depreciation, but also contribute in other residents' and passersby's sense of insecurity. The lack of sufficient attention to social behaviour of the immigrant population living collectively in cramped houses leads to an even more reduced sense of security. Consequently, not only have the emigration of indigenous people have moved out, the replacement residents have not helped the sense of desirability and security of the area (Fig. 16.4). Some characteristics of the new arrivals include the following:

(a) They are young and optimistic about their future income, and consider this mode of residency impermanent and provisional. So, they are prepared to live with these tough conditions to provide more money to save for their family.
(b) In order to save more money, they are inclined to dwell with other single men in the same place.
(c) They do not have a sense of belonging to their neighbourhood because they cannot afford to buy a dwelling, and even feel no need to provide a permanent residence. To them, home means just a place for temporary living.

Fig. 16.4 Historical buildings deterioration; the main result of collective residence of young and single immigrant men

These new occupants are, therefore, people who:

- have a temporary view about their work, and life,
- are short-termists,
- are prepared to live in the worst conditions,
- do not feel any place attachment to the neighbourhood and even their homes,
- are only in touch with people of similar social standing,
- are isolated and outcast with no relation with the neighbours.
- suffer from xenophobia (especially against Afghans).
- worst, they consider these conditions as normal and inevitable.

However, as specialists we suppose they could behave like indigenous locals, take civic responsibilities and play significant and active roles in regeneration of the fabric where the native wealthy residents and even the national and local authorities are unable to prevent its destruction and deterioration.

Along with young, single immigrant men, there are also those who live in Sang-e Siah with their family, including those who have inherited their homes from their parents (often homeowners). Generally, old residents with sufficient income have moved from this district to new neighbourhoods of Shiraz, but the ownership of inherited properties are shared between several families. This often makes it difficult for heirs to renovate or sell their property and move on. So, they have to remain there. In other words, not only do some inheritance laws provide burdens to redevelopment of old structures, they also cause ownership conflicts and contribute to further depreciation of buildings.

Due to fear of illegal occupation by others, the heirs are forced to live in old houses without due regard to essential maintenance. The lack of the sense of place attachment amongst this group is associated with fading collective memories and longing for the character or identity of this region.

Despite the fact that the above-mentioned groups are unlikely to participate in the urban fabric regeneration, there is another group who have chosen to live in the old district of the city and are the backbone of the social capital of the region (Fig. 16.5). Although they can afford moving costs they prefer to remain in the area, thanks to its values, special character and identity. This group is primarily made of educated people who adhere to the memories, but also elderly couples with roots in this region.

There is, however, another group with a rather repulsive effect on the community, namely drug addicts and dealers. They are the group behind the lack of a sense of safety, security and consequently the destruction of the neighbourhood image. Hidden and indefensible spaces in demolished and dilapidated buildings provide appropriate space for their activities (Figs. 16.6 and 16.7).

Besides, the formal statistics demonstrate a marked decline in the number of female population and significant rise in male population leading to a masculine environment with a reduced sense of safety for women as well as a presence of anti-family trends (Fig. 16.8). Consequently, the number of family residents is in decline.

Fig. 16.5 The house of Mohandesi Family, one of the famous family in the Shiraz City who live voluntarily in this region

Fig. 16.6 Addicts, the primary repulsive force in the community

Moreover, the study of active population or labour force indicates over than 66% of over 15 population are economically inactive and dependent on others. High unemployment rates are considered as one of the main reasons for the lack of a sense of safety rooted in financial hardship and thereby and illegal jobs.

16 Challenges of Participatory Urban Design … 257

Fig. 16.7 Indefensible spaces in Sang-e Siah fabric as a result of abandon the old houses

Fig. 16.8 The high proportion of elderly to young people

16.4 Conclusion and Recommendations

As the above analysis demonstrates the major burdens of public participation in an urban project are the mental constraints of both residents and authorities, embodied in the misunderstandings of the participatory approaches, and challenging socio-economic condition of the community embodied in significant levels of social problems.

A number of measures should be taken to alleviate and eradicate these obstacles. To tackle the residents' and local authorities' mental constrains changes in overall structure of the society is needed. Iranian society has suffered a lack of self-steam rooted in misinterpretation of city modernisation which has led to destruction of solidarity and cooperation in the society. Bearing in mind the historically and traditionally significant differences between Western and Iranian societies, the adoption of strategies from developed countries would not be appropriate for addressing these complications. Below are recommendation as to how to tackle problems of such areas:

(a) Being realistic about the expectation of young single immigrant men; making efforts to replace this group by enthusiastic local families, to expect better participation levels regeneration.
(b) New laws to eliminate the uncertainties surrounding inherited properties.
(c) Legislation to facilitate the purchase or rent of valuable and historical properties by municipalities in order to create new consistent activities in old constructions (e.g. using sizeable historical houses for creating unique traditional hotels and restaurants).
(d) Efficient use of indigenous residents' intellectual resources to identify and maintain collective memories plays crucial role in promoting a sense of place attachment amongst this group. Not only do they know restrictions and obstructions of this historical district, they are also the best advocates and promoters of regeneration. Welcoming this group's tangible and achievable ideas would pave the way for participation by other groups. Using this group's ideas in Sang-e Siah district regarding the features of the old neighbourhood was the main reason for attracting notable investors. They act as social capitals and absorb donors' and cultured peoples' assistance as well as locals.
(e) Due regard to needs and expectations of religious minorities as well as their collective memories.
(f) The identification of drug addicts for tough and appropriate acts, and the removal of non-resident addicts and dealers.
(g) The adoption of policies encouraging the return of the families to the neighbourhood along with organising special events in public places of this region can improve the sense of safety amongst residents and visitors.

References

Carmona M, Heath T, Oc T, Tiesdel S (2003) Public places urban spaces. Routledge, Oxford

Lee Y (2008) Design participation tactics: the challenges and new roles for designers in the co-design process. CoDesign 4(1):31–50. https://doi.org/10.1080/15710880701875613

Linden A, Billingham J (1998) History of urban design group. In: Urban Design Group. Urban Design Source Book, UDG, Oxford, pp 40–170

Manzini E, Rizzo F (2011) Small projects/large changes: participatory design as an open participated process. CoDesign 7(3–4):199–215. https://doi.org/10.1080/15710882.2011.630472

Pakzad J (2002) Expert culture and people culture. Urban Manag J 08:32–41 (in Farsi)

Pakzad J (2004) Comparative study of Iranian and European cities in order to finding historical obstacles of public participation. Soffeh No. 37:25–41 (in Farsi)

Safavi A (2011) Investigating the role of urban development management system in good urban design implementation—case of Iran. Urban Manag J Winter Autumn 28:255–272 (in Farsi)

Saghafi Asl A, Zebardast E, Majedi H (2016) Evaluation of implementing urban design projects in Iran, case study: implemented projects in Tehran. Armanshahr Architect Urban Develop J 9(17):185–197 (in Farsi)

Sanders E, Stappers P (2008) Co-creation and the new landscapes of design. CoDesign 4(1):5–18

Shahid Beheshti University (2016) Pathological study of urban design projects in Iran. A report to Ministry of Roads and Urban Development (in Farsi)

Tan L, Szebeko D (2009). Co-designing for dementia: the Alzheimer 100 project. AMJ 1(12):185–198

Index

A
Aalen, 177
Abadan, 175–177, 179–182, 186, 189, 191
Abgineh, 218
Addaraj, 138
Adjacent parcels, 58
Adriatic sea, 98
Afchangi house, Sabzevar, 159 (fig.11.1), 162 (fig.11.3), 163 (fig.11.4)
Affordable housing, 193, 194, 197, 200, 201, 203
Agha Mohammad Khan, 208
Ahari, 176, 178
Ahwaz, 176
Akhi, 108, 109
Al-ajir, 109
Alborz, 225
Aldaqi house, Sabzevar, 159 (fig. 11.1), 161 (fig. 11.2)
Aleksic, 93, 99
Aleppo, 107, 117, 118
Al Fateh, 39
Al-haush, 79
Aliados, 237, 241
Alleyways, 124–126, 128
Almasieh, 29
Almoghany, 139
Al-mu'allim, 109
Almubaed, 138
Al-mubtadi, 109
Al-Muhtasib, 108
Al Muqaddimah, 111
Al-qadi, 109
Al-Qasimi, *see* Mohammad Said al-Qasimi, 110, 111
Alramlawi, 4
Al-sani, 109
Al-shaykh, 109
Amini, 232
Amiri (Sadidi) house, Sabzevar, 159 (fig.11.1)
Andalib, 62
Andaruni, 163
Anderson, 12
Arab, 78, 79, 82
Architecture, 122, 128
Ardalan, 182
Armenian, 26
Arrangement and support of housing provision and implementation law, 59
ARS Progetti, 80, 82
Asgari, 65
Ashajaiah, 138
Ashoura, 222
Asia, 93, 96, 97
Attofah, 138
Austria, 96
Authorities, 247, 248, 251–253, 255, 258
Awareness-raising, 64
Ayyubid, 109
Azimian house, Sabzevar, 159 (fig. 11.1), 166, 167 (fig. 11.5), 168 (fig. 11.6), 170 (fig. 11.7)
Azzaitoon, 138

B
Baa'th, 128, 130
Baa'thist, 123, 128
Bab Al-Bahrain, 9, 13, 15, 16, 18
Bab-e Homayoun, 29
Babylonian, 129, 130

Badland, 230
Baghdad, 128
Baharestan, 29
Bahrain, 9–13, 15, 31, 32, 38, 40, 48
Bahrainy, 230
Bailey, 61
Baker Festival, 26
Baladiyat, 110
Balkan countries, 97
Balkan region, 96
Bam, 200
Banat, 98
Banister, 231
Banyn, 230
Baqani house, Sabzevar, 159 (fig. 11.1), 162 (fig. 11.3), 169, 171 (fig. 11.10)
Bartolini, 24
Bastay, 88
Battisti, 231
Bazaar, 211, 216, 218, 220, 225, 238, 241
Behjat Abad, 211, 218, 225
Bela Crkva, 98
Ben-Hamouche, 10
Bento, S., 241
Biruni, 163
Blackmore, 95
Bolaghi, 241
Bolhão, 237
Boshroye, 171 (fig. 11.6)
Bosnia, 99
Bou Ali Sina, 232
Boverdeh, 175
Bovista, 232
Braim, 175, 176, 179, 181–183, 185–187, 191
Buccolieri, 230
Buffer Zone, 75, 76, 80, 81, 83, 86, 89
Bukovina, 98
Bulgaria, 96
Byzantine, 95, 97

C

Cabinet of Iran, 59
Cambra, 231
Cameron, S., 196, 197
Canton, 96
Casa de Música, 232
Central silk management, 100
Cervero, 230
Chahar-bagh, 21
Chahar-soffeh, 166, 168, 168 (fig. 11.6), 169 (fig. 11.7), 170 (fig. 11.8, fig. 11.9)
Champs Elyseé, 29
Chandni Chowk, 27, 28
Chardin, 26

Cheshomi house, Sabzevar, 159 (fig. 11.1)
China, 94–97, 99
Choi, 236
Chragh, 81
Chu Ka Cheong, 141
Citadel, 121, 124–131, 134
Citadel, also Qala, 75–77, 80, 81, 83, 87, 88
Coercive possession, 58, 72
Commodified housing, 193, 203
Communal activities, 247
Communities, 248
Community-based organisations, 64, 72
Community organisations, 64
Compartmentalised approach, 69
Comprehensive plan, 60, 65
Conflict, 122–124, 130
Construction permissions, 65, 67
Courtyard, 124–126, 128, 129
Creighton, 231
Crinson, 181
Croatia, 99
Croatian agricultural bank, 99
Cubukcu, 231
Cultural initiative, 248
Culture, 122, 123, 125, 128, 132–134
Cyprus, 22

D

Dallman, 229
Dalmatia, 99
Damascus, 107, 110–118
Dan, 231
Dar al-Handasa, 80
Dareyni house, Sabzevar, 159 (fig. 11.1)
Dayaratne, 11
Declining area, 193–195, 197, 200
De-commodification, 203
Deimary, 168 (fig. 11.6)
Delal Khaneh bazaar, 81, 85
Delhi, 21, 28
Delhi Durbar, 28
Denniss, 60
DepthmapX, 235
Deputy of Urban Planning and Architecture, 65
Derrible, 231
Deteriorated areas, 57, 67
Deteriorated block, 57
Deteriorated neighbourhood, 56, 58, 60, 62, 63, 65–67, 72
Dezful, 189
Di Sabatino, 229
District municipalities, 66, 69, 72
Diwakhana, 124
Djukic, 102, 103

Index 263

Don Garcia de Figueroa, 26
Douro, 232
Downtown Erbil shopping centre, 81
Drab-Kooshek, 241
Drouville, 161
Duque, 231

E
Educational initiative, 248
Elias Qoudsi, 111
Enqelab, 218
Environmental Impact Assessment (EIA), 214
Environmental qualities, 248
Erbil, 121–134
Erbil, also Arbil and Hawler, 75–80, 82, 84
Erbil Master Plan, 80
Eslami house, Sabzevar, 159 (fig. 11.1), 161, 162 (fig. 11.3), 163 (fig. 11.4), 171
Esnaf, 108
Estaji, 159–163, 167, 170
Etzioni, 60
Euphrates, 179
Europe, 93, 95–97, 99, 100

F
Facilitation, 55, 59, 61, 62–65, 67, 69–72
Facilitation office, 55, 59, 62, 64, 72
Facilitative approach, 58
Fallahzadegan, 61
Farrell, 230
Federico, 94–96
Félix, 138
Ferdowsi, 238
Ferreira, 230
Fonseca, 229, 231
Fornara, 230
Foroughmand Araabi, 6
Forsyth, 231
France, 95
French industry, 96
Freud, 24
Fuccaro, 10

G
Garbero, 230
Garrick, 231
Gaza, 137–139, 141, 145, 146, 153
Gedik, 109
Gentrification, 196, 197, 200, 202, 203
Geographic Information System (GIS), 211, 220, 230, 236
Ghavampoor, 61
Gigi, 231

Gilderbloom, 230
Global North, The, 197, 198
Goblot, 208
Going for Growth, 196
Golestan, 211, 216, 218, 220
Gonçalves, 231
Good, 95, 96
Gospodini, 230
Grajewski, 231
Grecu, 230
Greece, 95
Groundworks, 64
Gulf, 10

H
Haghighi, 176, 180
Hajialiakbari, 56, 72
Halbwachs, 26
Hamad ben Khalefa, 40
Hamedan, 59
Hangzhou Silk Museum, 94
Hansen, 94, 95
Hanson, 231
Hatami, 232
Haveli, 28
Hawking, 17
Hejazi house, Sabzevar, 159 (fig. 11.1), 162 (fig. 11.3)
Henan Province, 95
Hepguzel, 231
Heritage Impact Assessments (HIA), 207, 210, 214, 216
Herzegovina, 99
High Advisory Council, 109
High Commission for Erbil Citadel, 79
High Commission for Erbil Citadel Revitalisation, 77, 123
Hillier, 231
Hirfa, 108
Historical district, 247, 248, 250, 253, 258
Historic centre, 211, 212
Historic Landscape, 216, 218, 220, 222, 225
Historic Urban Landscape (HUL), 23–25, 207, 210, 213, 216, 224
Hogan, 61
Hokm Abad, 232
Holden, 35
Holton, 12
Housing, 124–126, 131
Housing-led approach, 193–196, 201–203
Housing market, 194, 196, 199–203
Housing reconstruction, 55, 65, 70, 73
Howard, 177

Hunainiyah, 9, 13, 16, 17
Hunter, 61
Hydrocity, 211

I
Ibn Khaldun, 111
ICHTO, 171 (fig. 11.10)
ICOMOS, 210
Iida, 231
Imam Khomeini, 218, 220
Implementation, 247, 248
India, 15, 95
Indian, 26, 95, 96
Informal settlement neighbourhoods, 59
Intangible Cultural heritage, 216, 222
Interview, 247, 252
Iran, 175, 176, 180, 182, 194, 198, 200, 201, 203, 207, 208, 214, 216, 247, 248, 253
Iraq, 121–124, 128
Isfahan, 21, 59
Islamic, 22
Istanbul, 84
Iwan, 124
IWAN Centre, 138, 141, 143, 146, 147

J
Jabbari, 231
Jahanara Begum, 27
Jajar-zadeh house, Sabzevar, 159 (fig. 11.1)
Jamal al Din al Qasimi, 110
Jamei, 229
Japan, 95, 99
Japanese International Cooperation Agency (JICA), The, 57
Jayasinghe, 231, 236
Jeong, 230, 236
Jew, 26
Jiang, 231
Jigyasu, 145
Johansson, 230
Jomhouri, 218
Jones, 12
Joobare neighbourhood rehabilitation, 59
Joolan neighbourhood renewal, 59
Júlio Dinis, 232

K
Kalfa, 112, 118
Karimi, 229
Karim Khan, 218
Karun, 179
Kasemsri, 231
Kashgar, 95
Kasmai, 176

Keall, 166
Kermani-Moqaddam, 171 (fig. 11.10)
Khailfa, 112
Khalil Al Azem, 110
Khanaqa, 78, 82
Khans, 78
Kheirabadi, 160
Khiliu, 109
Khirokita, 22
Khoob-Bakht, 62, 63
Khorramshahr, 176, 177
Kian house, Sabzevar, 159 (fig. 11.1), 161
Kingdom of Serbs, Croats and Slovenes, the, 98
Klein, 230
Kockelman, 230
Kocovic, 102, 103
Koh, 231
Korea, 95
Kosovo, 98
Kuchas, 28
Kurdish, 121–125, 129–132
Kurdistan, 123, 130, 132
Kursi, 162

L
Labour government, 60
Lahore, 27, 28
Laleh Zar, 218
Land readjustment, 61, 65, 66
Land Readjustment Projects (LRPs), 58, 62, 65, 67, 69, 70–72
Lapovo, 99
Lawless, P., 198
Laylavi, 64
Lee, 231
Leide, 111
Lerman, 230
Leung, 150
Li, 230
Liu, 231
Liversedge, 35
Local identity, 122
Local office, 58
Long-term policy, 248
Long-term programmes, 64
López-Domínguez, 230
Lord Hardings, 28
Louis Massignon, 110
Loulan, 95
Low-interest loans, 65
Lundberg, 231
Lupton, R., 197, 198
Lyon, 96, 100

M

Macedonia, 99
Maddison, 60
Maginn, 60
Mahallas, 77
Mahdavi, 160, 162, 164
Mahdavinejad, 232
Makedonija, 101
Mamluk, 109
Manama Port, 10
Market failure, 203
Markovic, 94, 96, 97, 102
Marquet, 231
Martinelli, 231
Mashhadi house, Sabzevar, 159 (fig. 11.1), 161, 163 (fig. 11.4)
Matzarakis, 231
McCahill, 231
Meares, 231
Mediterranean, 95–97
Medumurje, 99
Mehrgerd, 207, 209–211, 214, 216, 218, 220, 225
Mehta, 231
Meidan, 114
Mental constrains, 258
Merkel, 95
Mid-term policy, 248
Mid-term projects, 64
Milan, 96, 100
Military Housing Corporation, 118
Mimar bashi, 112
Ministry for Agriculture and Water (MAW), the, 100
Ministry of Housing and Urban Planning, 59
Ministry of Public Works and Housing,, 140
Ministry of Trade and Industry (MTI), The, 100
Miralles-Guasch, 231
Mohammadiyani house, Sabzevar, 159 (fig. 11.1)
Mohammad Said al-Qasimi, 110
Molavi, 232
Montenegro, 99
Moonlight Square, 27
Morar, 230
Morphological analysis, 121, 123, 128, 131
Moslem house, Sabzevar, 159 (fig. 11.1)
Mostowfi, 161
Moura, 231
Mouzinho da Silveira, 241
Mu'allim, 112
Mughal, 28
Muhaisen, 145

Muhaisen and Alheta, 145
Muharraq, 16, 31, 32, 34, 38–42, 51
Mukhtar, 79, 83
Mulberry trees, 93, 94, 98–100
Multi-Criteria Analysis (MCA), 230, 238
Municipality of Tehran, 57
Murayama, 61
Murphy, 34

N

Naderi, 238
Nadir Shah, 28
Nafisi, 163
Nahavandi, 231
Naqsh-e Jahan, 26, 27
Naser-eddin Shah, 29
Nasir, 231
National Garden, 216
National Library in Belgrade, 100
Navvab, 199
Navvab Highway, 58
Neighborhood, 128, 129, 134, 247, 249, 250, 252, 254, 255, 258
Neighbourhood organisations, 61, 62, 64, 72
Neighbourhood renewal, 57, 60–62, 72, 73
Neighbourhood renewal plan, 64, 66, 69, 72, 73
Neighbourhood stakeholders, 69
Nishiuchi, 231
Nishtiman bazaar, 81
Nova Kanjiza city, 101
Novi Sad, 99
Nubdha tarikhiyya fil hiraf al Dimashqiyya, 111
Nvivo, 13

O

Oakes, 231
Obstacles of participation, 247
Office of the Deputy Prime Minister, 60
Omer, 231
Önder, 231
Onley, 10
Ottoman, 108–110, 113, 114, 117

P

Pahlavi period, 159, 159 (fig.11.1), 160, 162, 163, 163 (fig.11.4), 171
Pakzad, 248
Palácio de Cristal, 232
Palestine, 138
Panagopoulos, 231
Panahi house, Boshroye, 159 (fig.11.6)
Pancevo, 101

Panchevo, 98
Park Shahr, 218
Parliment of Iran, 65
Parthians, 166
Participation, 247–249, 252, 253, 258
Participatory activities, 248
Participatory approaches, 247, 249, 252, 258
Participatory urban design projects, 247, 248
Partnership contract, 65
Pearling Trail, 9, 13, 16
Pec Patriarchate, 98
Pedestrian Network Assessment (PNA), 230
Peiravian, 231
Pelegrino, 99
Penn, 231
Peponis, 231
Perm, 231
Persia, 95
Persian, 21
Persian Garden, 220
Physical Development Institute, 56
Pierson, 56
Pietro Della Valle, 26
Place-identity, 121–123, 129–134
Planning, 123, 124, 128, 129, 133, 134
Polycentric, 230, 232, 241
Population density, 56
Porter, L., 195–197, 202
Portugal (Porto), 229–238, 240–242
Post-war, 122, 130, 131, 133, 134
Pourjafar, 232
Poverty, 193–197, 199
Powell, 11
Praça da Liberdade, 241
Private investors, 64, 73
Problematic procedure, 247, 248
Procedure, 252, 253
Public authorities, 69
Public participation, 247–249, 252, 253, 258
Public sector agencies, 64, 65
Public services provision, 65
Public Silk Factories, 100
Public space rehabilitation, 66

Q
Qajar, 21, 28, 29, 208
Qajar period, 159 (fig.11.1), 160, 162, 163 (fig.11.4)
Qala, also Citadel, 77, 78, 88
Qamus al-ṣinaʿat al-Shamiyya, 110
Qanat, 207
Qanat, 207–211, 216, 218, 220, 225
Qaysary Bazaar, also Qaysary Suq, 78
Qaysary Suq, also Qaysary Bazaar, 78, 81

Qazvin, 229–233, 235–238, 240, 241
Qualitative interview, 247
Quality of life, 195, 196

R
Rahimieh, 178
Raith, 160, 170, 172
Rajagopalan, 229
Ramos, 231
Rasht, 59
Reconstruction, 121–123, 130–135
Red Cross, 137, 138, 146
Red Flower Festival, 26
Red Fort, 27, 28
Regeneration, 193–196, 198–204
René Danger, 114
Renewal plans, 62
Renewal process, 61, 66, 69, 72, 73
Renewal service offices, 59
Rent-speculation, 199
Residents' perception, 247, 252
Revitalisation, 123
Riggs, 231
Riley, 12
Rishbeth, 11
Roboobi, 178
Rofè, 231
Rogac Mijatovic, 102
Rome, 95
Rostampour, 181, 188

S
Sabatino, 229
Sabzeh Meidan, 218
Sabzevar, 157, 158, 159, 159 (fig.11.1), 172 (fig.11.6), 173
Sacrifice Holliday, 26
Safavid, 21, 25, 26, 28
Safavid period, 159 (fig.11.1)
Salehi, 61
Salizzoni, 230
Sana'a, 16
Sang-e Siah, 247–250, 252–255, 257, 258
Sano, 231
Sanoff, 150
Santa Cantarina, 232
Sar-kocheh Reyhan, 241
Sassanians, 166
Schmitz, 231
Seljuk, 108
Serbia, 93, 97–100
Serbian heritage, 98
Sericulture in Serbia, 98
Sesic, 102

Shahid Ansari, 232
Shah Jahan, 27
Shahjahanabad, 27
Shanasheel, 79
Shanghai, 95
Sharan Engineering Consultants, 56
Shar Square, 81, 82
Shaw, K., 195
Shaykh al-mashayekh, 109
Shekalla bazaar, 81
Shi'a, 26
Shiraz, 59, 247, 248, 250, 252, 253, 255, 256
Short-term activities, 64
Shushtar New Town, 181
Sicily, 95
Silk Roads, 93, 95, 96
Sinf, 108
Slavonia, 99
Socharoentum, 229, 231
Social deprivation, 193
Social exclusion, 56, 193
Social Exclusion Unit, 60
Social groups, 247, 248
Socially rooted problems, 247
Social obstacles, 247
Social problems, 247, 252, 258
Socio-economic characteristics, 248
Socio-economic conditions, 247, 252, 253, 258
Socio-spatial pattern, 193
Souk, 9, 13, 15
Soulhac, 230
Souri, 6
Space Syntax, 229, 230, 235–237
Spain, 95
Stakeholders, 247–249
State Department of sericulture, 99
State-led, 203
Sternudd, 230
Street Connectivity Analysis (SCN), 230
Supreme Council of Architecture and Urban Planning of Iran, 57
Suq, 78
Surplus density, 65

T
Tabriz, 59
Tā'ifa, 108
Tajeel, 78, 79, 82, 86–88, 90
Taklamakan, 95
Tanzimat, 110
Tavassoli, 232
Ta'zīye, 222

Tehran, 21, 29, 55–62, 63–67, 69, 72, 193, 195, 198–203, 207–211, 216, 217, 222, 225
Tehran City Council, 65
Tehran Grand Bazaar, 216
Tell, 77, 78
Temporary habitation, 65
Temporary settlement, 62
Tigris, 179
Timurid, 21
Timurid period, 159 (fig.11.1), 166
Titov Veles, 101
Toopkhaneh, 29
Towchal, 208
Towhidi-Manesh, 167 (fig.11.5)
Tradition, 122, 123, 128, 134, 135
Tumer, 231
Tunstall, R., 197

U
UK, 96
UNESCO, 9, 13, 23, 25, 76, 79, 80, 82, 207, 210, 212, 214
UNESCO World Heritage Committee, 214
UN-Habitat, 77
Urban design, 121, 123, 128, 134
Urban design projects, 247, 248, 253
Urban deterioration, 56, 59, 65
Urban Development and Rehabilitation Corporation (UDRC), 59
Urban development plans, 69
Urban green infrastructures, 207, 222
Urban heritage, 207, 212, 222
Urban Renewal Organisation of Tehran, 57
Urban space, 68, 72

V
Vadoodi, 61
Valiasr, 218
Varzaneh, 232
Vojvodina, 99, 100

W
Wang and Li, 141
Weber, 231
Weighted Linear Combination (WLC), 230, 234
Welfare, 201, 203
Welfare system, 203
Welwyn Garden City, 179
Willis, 230
Wong, 231

World Heritage (WH), 210–212, 214, 216, 222
World Heritage Site, 75, 76, 80, 121, 124, 130, 131

X
Xiao, 230
Xu, 230

Y
Yangtze River, 94
Yarjani, 225

Yau, 141
Ye, 230
Yugoslavia, 96–102

Z
Zadeh, 231
Zartosht, 211
Zavvare, 168 (fig.11.6)
Zhou, 236
Zoroastrians, 26

Printed by Printforce, the Netherlands